W9-AAR-062

COUNTDOWN TO
APOCALYPSE

Other Books by Paul Halpern

The Pursuit of Destiny

The Quest for Alien Planets

The Structure of the Universe
(with Bruce Gregory)

The Cyclical Serpent
(with Felicia Hurewitz)

COUNTDOWN TO APOCALYPSE

A Scientific Exploration of the End of the World

PAUL HALPERN

PERSEUS PUBLISHING

Cambridge, Massachusetts

Many of the designations used by manufacturers and sellers to distinguish their products are claimed as trademarks. Where those designations appear in this book and Perseus Publishing was aware of a trademark claim, the designations have been printed in initial capital letters.

Copright © 1998 by Paul Halpern

All rights reserved. No part of this publication may be reproduced, stored in a retrieval system, or transmitted, in any form or by any means, electronic, mechanical, photocopying, recording, or otherwise, without the prior written permission of the publisher. Printed in the United States of America.

Library of Congress Catalog Card Number: 00-105210
ISBN 0-7382-0358-0

Perseus Publishing is a member of the Perseus Books Group.

Find us on the World Wide Web at http://www.perseuspublishing.com

Perseus Publishing books are available at special discounts for bulk purchases in the U.S. by corporations, institutions, and other organizations. For more information, please contact the Special Markets Department at HarperCollins Publishers, 10 East 53rd Street, New York, NY 10022, or call 1-212-207-7528.

First paperback printing, July 2000
1 2 3 4 5 6 7 8 9 10—03 02 01 00

For Felicia, my joy and inspiration

CONTENTS

PREFACE

TERRORS OF THE NEW MILLENNIUM

> *Some say the world will end in fire*
> *some say in ice . . .*
>
> —Robert Frost, "Fire and Ice"

How will the world end? In intense heat or in extreme cold? Will it happen in a flash, or will it take centuries? Might the impact of a comet bring it about—or perhaps a powerful nuclear explosion? For thousand of years, sages and scientists, poets and prophets alike have been speculating about what will occur when our time on Earth draws to a close. At the dawn of the third millennium it is only fitting to re-examine the question: What will the final days of our world be like?

The world is passing through an invisible curtain of time, emblematic of an age of profound transition. On one side of the curtain lies the tested and familiar—the art, philosophy, religion, and science of the second millennium A.D. On the other side stands the unknown—the history and culture of the next era.

We have left behind a millennium of phenomenal progress. The social condition of the human race has changed so much in the past thousand years. As a species, our political and economic growth has been amazing. We have evolved from a patchwork arrangement of city-states, farming communities and nomadic tribes to a highly complex network of multinational organizations. No longer can large segments of the globe remain isolated and indifferent to the world at large.

Meanwhile, international exchange of products and ideas has made possible the blossoming of technology on an unprecedented scale. New materials and sources of energy have been discovered, new tools for farming and construction have been fashioned, and new types of transportation and communication have been developed. From the clipper ship to the airplane, from the mechanical clock to the digital computer, and from paper money to the credit card, these innovations have altered the face of the globe. In short, there has been a veritable revolution in the way people conduct their lives—both in their homes and workplaces.

Sadly, parallel to this growth in the constructive use of technology has come a corresponding increase in the destructive power of weaponry. Over the past centuries, swords, spears, and shields have been replaced by mines, missiles, and machine guns. The capacity for people to destroy each other has augmented more than a million-fold. It used to be battlefield causalities numbered in the tens or hundreds, the result of prolonged hand-to-handcombat. Now it is possible for leaders to sit in their offices and order the instant annihilation of millions of civilians.

No wonder this new era engenders such trepidation. Will the third millennium A.D. be an era of continued progress, increased prosperity and movement toward global peace? Or will it represent the closing chapter of a long history of brutality? Will it constitute a time of renewal or an age of corruption, a fresh start or a final straw?

Why did the arrival of the year 2000, in particular, draw so much interest? Maybe it was because of its numerological significance: the first time in the modern era that the year was divisible by one thousand. Or perhaps with its enigmatic assembly of three zeroes in a row, "2000" seemed like a blank slate—a tablet upon which our hopes and fears of the future could be written. Now that the anticipated year has arrived, we shudder as we stand on its threshold.

For centuries, essayists and novelists alike pondered how civilization would appear at the close of the second millennium. Often, these writers were quite optimistic in their predictions. Based on the tremendous advances of the past centuries, many future-directed thinkers projected the year 2000 to be a time of far-reaching technological progress. They pictured it to be an era of untold marvels, unsurpassed riches, and abundant leisure. Some thinkers, such as influential 19th century American author Edward Bellamy, even imagined that social development would be such that by now the human race would exist in a state of utopia. As recently as 1966 an article appeared in Time magazine reporting that, "By 2000, machines will be producing so much that everyone in the U.S. will, in effect, be independently wealthy . . . How to use leisure meaningfully will be a major problem and [one expert] foresees a pleasure-oriented society full of 'wholesome degeneracy . . .' According to one estimate, only 10 percent of the population will be working and the rest will, in effect, have to be paid to remain idle."[1]

By no means have all prognostications of the third millennium been positive. Pessimists often use the brutal facts of the 20th century to prove that the human race is rapidly heading toward doom. Because of deep flaws in the human psyche—such as incessant greed, a propensity for violence, and lack of environmental concern—these pessimists feel that civilization is heading for certain destruction. Because warfare has grown more vicious and weaponry more deadly in the past decades, these naysayers conclude that armed conflict will expand even further in the years to come. Because pollution has taken a horrendous toll so far, they project that it will soon devastate the globe. Because poverty and social inequality have resisted attempts at amelioration, they assume that these conditions are beyond solution.

The extreme statement of this bleak point of view is a belief that global apocalypse is imminent. As the third millennium draws near, apocalyptic cultists have increasingly surfaced. Sects such as the Peoples Temple in Guyana, Order of the Solar Temple in Switzerland and France, the Branch Davidians in Waco, Texas, Heaven's Gate in California, and the Aum Shinrikyo cult in Japan have used dire warnings of doom to attract panicked followers to their flock. In many cases the anxiety experienced by the members of these groups has been exploited by charismatic leaders. Cultists have been manipulated by their fears, and cajoled to perform extreme actions such as homicide or suicide. Typically, they hope that these radical measures will help them pass smoothly into new realms beyond what they see as a dying Earth. Shocking events such as the mass murder-suicide of 913 followers of Jim Jones in the late 1970s, the mass murder-suicide of fifty-three members of the Order of the Solar Temple in 1994 (followed by additional deaths of Solar Temple members in 1996 and 1997), the release of the nerve gas sarin into the Tokyo subway system by members of Aum Shinrikyo in 1995, and the ritual mass suicide of thirty-nine members of Heaven's Gate in 1997, are emblematic of this panic in the face of imminent earthly disaster.

Why has the arrival of the millennium triggered such a strong response among doomsday sects? Perhaps it is because of this period's powerful connotations in Christian theology. Apocalyptic sects often use New Testament imagery to justify their belief that the "end days" are near. Sharing little in spirit or substance with mainstream churches, they nevertheless exploit the nuances of the Bible to justify their own catastrophic goals and desires.

The popular media also reflect a broad current interest in apocalyptic ideas. Recent films such as *Armageddon* and *Deep Impact*, as well as television shows such as "Millennium," indicate the public's deep fascination with how the world might end. The tremendous anxiety associated with the "Y2K problem" connected with the coming of 2000 provides yet another example of such trepidation.

Despite the colloquial use of the term today to describe a wide range of calamities, apocalyptic thought began exclusively as a set of religious notions. Apocalypse, a word of Greek origins, literally means "unveiling." It was originally used solely to refer to the Revelation to John, the last book of the New Testament, and other predictive writings from the early Christian period. In practice, it has come to represent the period of colossal devastation prophesied to occur before the Second Coming of Jesus. Over the ages, it has also come to designate eras of universal catastrophe, or even times of great natural disaster, such as extensive floods or fires. A major branch of theology, called *eschatology*, from the Greek "eschatos," meaning "last," concerns itself with varied interpretations of God's revelation of the future of Earth and the cosmos.

According to most Christian interpretations, the biblical apocalypse is a prophesied future era ordained by God to bring Earth's history to a close. The New Testament foretells that it will begin when Jesus, in heaven, opens the seven seals of a scroll upon which is written the destiny of humankind. As each of the first four seals are opened, a horse and rider will prepare for journey to Earth. When these four riders, known as the four horsemen of the apocalypse, reach Earth they will usher in a great wave of destruction. Only the pure of heart and body—the servants of God—will be spared from this mayhem.

The book of Revelation is quite vivid in its description of this prophesied age of chaos. As angels blow ominous trumpets, a host of tragedies will be inflicted upon humankind. These will include (listed here in no particular order) a rain of hail and fire, a plague of locusts, the falling of stars from the sky, the instant annihilation of the creatures living in the sea, the pollution of the oceans with blood, and the painful extinction of one-third of the human race. Only after these tribulations finish will Jesus come and rule the world in an age of peace and harmony.

At the end of Revelation appears the admonition by John, the book's author, "I warn everyone who hears the words of the prophecy of this book: if anyone adds to them, God will add to him the plagues described in this book, and if anyone takes away from the words of the book of this prophecy, God will take away his share in the tree of life and in the holy city, which are described in this book."[2]

Nevertheless, in spite of this dire warning, some contemporary Christians (though certainly not all) interpret Revelation in a less-than-literal manner. Rather than considering it to be an exact account of how humankind's time on Earth will draw to a close, they instead view the text as an allegory of the choices that lie ahead for our race. If global apocalypse were to occur, they feel, it would likely be the direct result of human shortsightedness and stupidity, rather than the inevitable outcome of God's edicts. It would be carried out by the launching of atomic bombs, chemical missiles, biological warfare devices, or other weaponry of the sort, rather than literally by four horsemen. And if humankind were to choose peace instead of war then perhaps global apocalypse could be avoided altogether, or at least postponed until the human race meets a more natural demise.

It is in this figurative sense that science considers the possibility of apocalypse. Though scientists often borrow nomenclature and imagery from the biblical account, they use them for purely secular purposes. They do so because the images from the book of Revelation are so evocative that these pictures help us to fathom better the scope of possible global disaster. Thus, scientists freely employ expressions such as "nuclear armageddon" and "ecological apocalypse" in order to emphasize the horrors of the fate that we may someday bring upon ourselves. (The biblical term "armageddon" refers to a site where the ultimate battle between the great kingdoms of the world is prophesied to be fought.)

We are fortunate to be creatures of free will, with the ability to affect our fate as a species. Now, more than ever before, we possess the power to reshape our surroundings—to mold our planet to suit our interests. Through the choices we make, we could steer our planet toward paradise or transform it into hell on Earth. In the dawn of the third millennium, we'd best take stock of ourselves and develop strategies to preserve our fragile environment. Moreover, we must try to eliminate warfare in all of its forms: nuclear, chemical, biological and conventional. How quickly we take these steps may decide if the year 2000 represents the beginning of a new lease on life for our people or, rather, simply the beginning of the end.

Sadly, though, in spite of our best efforts, the history of the world will come to a close someday. There is only so much we can do to postpone the inevitable. Avoiding devastating conflict and reducing hazardous pollution may buy our civilization hundreds, thousands, maybe even millions of years. But sooner or later we will succumb to natural disasters beyond our control—a great plague for which there is no cure, an ecological catastrophe of lethal proportions, or the death of the Sun, for example. Perhaps we could avoid terrestrial demise by developing the means to escape to other planets. Even so, in billions of years we would ultimately face the end of the cosmos itself.

One hopes that when ultimate extinction finally arrives, the human race will end its days (assuming that there even is time to prepare) like the proud captain of a besieged ship—with a dignified fighting spirit, rather than with mere resignation or panic. As Dylan Thomas wrote, we must "rage against the dying of the light,"[3] but we must do so with presence of mind and composure, not with mass hysteria. Human history, a sacred volume, deserves to be bound in a suitably elegant cover.

ACKNOWLEDGMENTS

The guises of doom are many. To attempt to uncloak its hidden form, I have drawn on numerous fields of the natural and social sciences, in many of which I am not an expert. For this reason, I am deeply in debt to those individuals who have generously helped me decipher these disciplines.

I would like to thank Saul Perlmutter, Alan Hale, and Tom Gehrels for their kind assistance with this project. The staff of my institution, the University of the Sciences in Philadelphia, including Phil Gerbino, Barbara Byrne, Charles Gibley, Nancy Cunningham, Elizabeth Bressi–Stoppe, Paul Angiolillo, David Kerrick, Mary Rafferty, Judith Kuchinsky, Salar Alsardary, Durai Sabapathi, and Carol Weiss, have provided invaluable help and support for my research. Thanks to my colleagues William Walker, Bernard Brunner, Ray Orzechowski, and Ruy Tchao for their considerable assistance in reading over and commenting on sections of my manuscript related to their respective fields.

I would also like to thank my family and friends for endless support during my days of apocalyptic fervor. I appreciate the insights into medieval history provided by Simone Zelitch. Many thanks also to Stanley, Bernice, Richard, Anita, Alan, Kenneth and Esther Halpern, Arlene and Joseph Finston, Fred Schuepfer, Pam Quick, Scott Veggeberg, Marcie Glicksman, Michael Erlich, Fran Sugarman, Debra DeRuyver, and Dan Tobocman for their assistance.

Special thanks to the editorial staff at Plenum, particularly Linda Greenspan Regan, Vanessa Tibbits, Barbara Merchant, and Robert Maged, for helping to bring this book into the world well before the clock of universal doom strikes (one hopes).

Above all, I'd like to thank my wife Felicia and my son Eli for their magnificent love, support, and guidance. Without their inestimable contribution, this project never would have reached completion.

INTRODUCTION

ELEGY TO EXTINCTION

When they walked our paths, their footsteps brought
thunder.
Now they are gone, and the soil is silent.
When they drank from our lakes, they robbed the nectar of
the rains.
Now they have left, and the waters are brimming . . .

—P. HALPERN, *"Ballad of the Survivors"*

Death of a Species

Would the world notice if the human race vanished? Or would it feel cool indifference? Might it even issue a great sigh of relief if our marauding, domineering kind, that has plundered Earth's resources for generations, were suddenly wiped out?

Tens of millions of years ago, a highly established, seemingly well-adapted species that dominated the Earth for eons completely and mysteriously vanished. Life indeed went on. Other creatures, particularly mammals, welcomed the extinction of the race of great thunder lizards by quickly filling the power void. Soon, global ecology functioned smoothly again—as if the colossal reptiles never existed.

The story of the death of the dinosaurs serves as a valuable lesson for the human race. Like the life fires of the dinosaurs, and numerous other species before our day, the human flame lies perilously exposed to the winds of extinction. If the gusts of doom ever blow in an inauspicious direction, humankind's candle could be snuffed out forever. As in the case of countless creatures that now lie inert, no shelter exists (at least currently) that could permanently protect our vital energies from eradication.

Whatever terrible calamity it was that served to wipe out the dinosaurs has been a matter of controversy for generations. Might it have been the devastating blow of a cosmic catastrophe—such as the impact of a large asteroid—that caused their demise, or another natural process much more gradual? Scholars churn through geological data, seeking the answer to this pivotal question.

No one would dispute, however, the sobering fact that during an era that lasted for 150 million years, myriads of these beasts once trampled through the mud—and now there are none. There but for fortune walks our kind as well.

The Age of the Thunder Beasts

During the Cretaceous period (roughly over 65 million years ago), the Earth was a savage and tumultuous place. Primitive subtropical and tropical rain forest blanketed much of the globe, its growth spurred by a climate that was especially warm and humid. In these forested regions, the land was thick with colossal ferns and dank conifers, drinking up as much moisture as they could possibly hold. Among the vegetation typical of this period were the Cycads, seed plants bearing large intricate leaves crowning thick trunks. The first flowering plants vied with them for a place in the sun, adding color to a landscape otherwise etched in greens and browns.

Burrowing their way through the nooks and crannies of the forest floor were our ancestors, the early mammals. There were few mammalian species at that time; the bulk of these creatures were small and ratlike. Their diminutive size rendered them well adapted to a lifestyle of foraging among the overgrown vegetation. Unable to fend for themselves in the open, they avoided the seas and skies, and spent most of their time hugging the ground and clutching at the trees.

Bordering these wildlands were aquatic domains: spacious oceans, shallow seas, and murky swamps. These watery reaches were brimming with millions of primordial types of fish, mollusks, echinoderms (starfish and sea-urchins, for example), and foraminifera (organisms with perforated shells), as well as other forms of maritime life. With their bountiful array of shapes and sizes, these sea creatures comprised much of the world's diverse species at that time. Most of these primitive life forms have died out since and are therefore unfamiliar to us today.

As murky as the seas were, the skies were often even hazier. In many parts of the world—India, for instance—huge volcanic eruptions spewed forth enormous quantities of ash into the atmosphere. Meanwhile, sinuous streams of molten red lava carved out ghostly portraits in the rocks, clashing vividly with the blackened dome above. As tired vistas crumbled under the weight of erosion, new terrain was slowly built up in their stead. The pressed land gave way to new configurations, the compliant seas shifted to fill the gaps, and the fiery sun of the Cretaceous heavens blazed down upon it all.

Reigning over this primitive kingdom were the undisputed rulers of land, sea and sky: the giant reptiles. There was scarcely a part of Earth that these thundering beasts (dinosaurs, pterosaurs, and other related species) didn't roam. And where they roamed, they flourished; where they flourished, they triumphed.

Imagine the nightmarish spectacle of gargantuan "lizards," trampling upon the muddy ground, wading through steamy swamps, soaring, swooping, snatching, and supping. The meat-eaters among them, such as the fiercesome Tyrannosaurus rex, tore their sabre-like teeth through any living flesh available, including the substantive hides of other dinosaurs. No creature on Earth could escape these awesome predators, save by hiding in a niche too small for the mammoth reptiles to discover. And *T. Rex* didn't even represent the largest of the saurians. The eight ton *Gigantosaurus carolinii*, recently found by Argentine paleontologists in the remote region of Patagonia, is thought to have been even taller.

The giant reptiles were far more intelligent than once given credit. At one time it was generally believed that these creatures were sluggish, stupid, and unconcerned about their young. Now scientists realize that many dinosaurs could move fairly quickly and stalk their prey. Moreover, there is ample evidence that at least some of them displayed nesting behavior similar to birds. In Mongolia, skeletons of Protoceratops, dinosaurs from the late Cretaceous period, have been found along with nests of eggs that contained preserved embryos. These indications of a nesting instinct suggest that dinosaurs, like birds or mammals, may have to some extent fostered their young.

Soaring high above the dinosaurs were their aerodynamically gifted cousins, the pterosaurs or "flying lizards." Like the dinosaurs, many pterosaurs were huge—some even the size of airplanes. The largest pterosaur, named Quetzalcoatlus for the Aztec winged serpent god, had a wing span of over thirty-six feet. In spite of their often immense proportions, though, pterosaurs had no trouble flying. Their bones were hollow, rendering them lighter than their dimensions would suggest. They navigated through the air by flapping elongated wings.

The great reptiles reigned for 140 million years. Why don't the dinosaurs and pterosaurs rule the Earth today? Why is it that aside from their descendants, the birds and lizards, there isn't a single living example of these creatures? These questions beg for a satisfactory solution. Truly, the disappearance of the giant reptiles represents one of the greatest mysteries of science.

Fossil records paint a curious picture of mass extinction at the close of the Cretaceous period, 65 million years ago. Paleontologists have identified skeletal and other evidence of various dinosaur species for tens of millions of years up until that time. Yet after the Cretaceous period, the record of these animals' existence simply stops. The great thunder beasts, undisputed masters of the Earth, simply vanished without a trace (except, of course, for their fossilized remains).

Moreover, the saurians were far from the only creatures to become extinct at that time. In the brief interval between the end of the Cretaceous and the beginning of the next geological period, the Tertiary, considerable fossil evidence suggests that close to half of all maritime species died out. Numerous plant species abruptly disappeared as well. Tertiary vegetation bore scarce resemblance to its pre-

ERA	PERIOD	MILLION YEARS AGO
CENEZOIC	QUATERNARY	< 1
CENEZOIC	TERTIARY	1 - 65
MESOZOIC	CRETACEOUS	65-140
MESOZOIC	JURASSIC	140-215
MESOZOIC	TRIASSIC	215-250
PALEOZOIC	PERMIAN	250-280
PALEOZOIC	CARBONIFEROUS	280-360
PALEOZOIC	DEVONIAN	360-400
PALEOZOIC	SILURIAN	400-435
PALEOZOIC	ORDOVICIAN	435-500
PALEOZOIC	CAMBRIAN	500-600
PRE-CAMBRIAN	PROTEROZOIC, ARCHEOZOIC, AZOIC	> 600

Earth's geological chronology.

decessors. All in all, fossil records depict a wiping out of about 60 percent of all known living species from the face of Earth.

Many theories, as noted before, have been advanced over the years as to why the dinosaurs vanished. One school of thought has held that they became too big to survive, and therefore collapsed under their own weight. Another has argued that they were just too dumb to compete with other species. A third has suggested that they died out because small mammals ate all of their eggs.

None of these proposed solutions hold water. First of all, dinosaurs are known to have been extremely successful creatures while they lived. They survived for tens of millions of years. During that time, they vanquished all competitors. Although some of them grew large, they never became too big to move. Therefore, why would they have suddenly become defeated by other species? And as to the third mentioned hypothesis, it would have been physically impossible for mammals to have eaten all of the saurian eggs. If they had, they would have died out themselves for lack of food. Furthermore, though these theories attempt to address the extinction of the great reptiles, they fail to account for the much more significant demise of countless types of marine organisms.

It has only been in recent years that a sensible solution to the dinosaur mystery has been proposed. This novel hypothesis—that of sudden extinction by the impact of a body from space—offers the tantalizing prospect of also explaining the disappearance of numerous other missing species.

The Alvarez Proposal

Until the early 1980s, American physicist Luis Alvarez was known almost exclusively for his work in tracking down subatomic particles. His accomplishments in this field were most substantial. In 1947, he used the Berkeley's linear accelerator to produce free protons of record energy levels. Then, in 1955, he designed a unique way of analyzing the frozen tracks left by particles in bubble chamber detectors (containers filled with liquid hydrogen that record the motions of particles). His design helped scientists to discover dozens of new subatomic bodies. Finally, in 1960, he used bubble chambers to discover "resonances"—extremely short-lived particles that can be created with accelerators and detected in the laboratory. For this last discovery, he won the Nobel Prize for Physics in 1968.

Many scientists, at that point, would have rested on their laurels. Alvarez had won the respect of physicists everywhere; he had the luxury of retiring with great honor. He did eventually step down from the official duties of his professorship and become professor emeritus at the University of California at Berkeley, but he hardly stopped working.

In the late 1970s Alvarez began a geological research collaboration that was to last until his death in 1988. He joined his son, the noted Berkeley geologist Walter Alvarez, as well as Berkeley researchers Frank Asaro and Helen Michel, in investigating the clay-filled boundary between the Cretaceous and Tertiary periods. An analysis of this subterranean juncture, known as the K/T stratum, was expected to yield clues as to the nature of geological and climatic changes during the time in question.

Remarkably, in the layer they investigated, the Berkeley team found a marked excess of the rare chemical element iridium. Iridium is virtually never seen close to the surface of the Earth. When our planet was formed, most of its iridium sank to its core. This heavy element is somewhat more commonly detected in meteorites (extraterrestrial rocks falling to Earth). Yet the group found unusual amounts of this material in the K/T boundary throughout the world, from Italy to New Zealand.

The Alvarezes and their colleagues calculated that the impact of an asteroid of roughly six miles in diameter (or a comet of somewhat larger dimensions) would be enough to produce the measured iridium excess. They postulated that such a body crashed into Earth some sixty-five million years ago. Their estimates indicated that the sudden impact of such a fireball would spew 100 trillion tons of dust into the air, just enough to produce the two-inch-layer of clay that forms the K/T boundary. Furthermore, it would carve out a crater of approximately 100 miles in diameter.

The Berkeley scientists immediately realized that their theory—now called the Alvarez hypothesis—would readily explain the mass extinctions at the end of the Cretaceous period. Immediately upon impact, the heat from the fireball would ignite a significant percentage of the world's forests, turning much of our planet's extant vegetation into soot. Moreover, the cloud of dust generated by the cosmic collision would fill the skies for many months, blocking the Sun's direct rays and shrouding the Earth in near-complete darkness. As temperatures plummeted to winter levels throughout the globe, photosynthesis would be greatly suppressed, resulting in many additional forms of plant life dying out. Large animals, such as the dinosaurs, dependent on abundant supplies of vegetation (either directly in the case of plant-eaters, or indirectly in the case of meat-eaters), would similarly be wiped out. Chemicals produced in the fireball would fall to the ground in the form of acid rain. This, in turn, would pollute the vegetation and the oceans, resulting in the speedy eradication of numerous species, including marine life. In short, the majority of the world's living species would become extinct.

Substantial resistance met the Alvarezes' ideas at first. Many scientists scoffed at the idea of sudden extinction of species by an extraterrestrial agent. Perhaps this hesitation was due in part to a general antipathy toward any theory even remotely resembling the bizarre notions of Dr. Immanuel Velikovsky.

Velikovsky had caused quite a stir in the 1950s and 1960s with the publication of a series of books purporting to explain the origins of biblical and mythological tales through astrophysics. In his most famous work, *Worlds in Collision*, he had claimed that many of these legends stemmed from near-Earth collisions by the planets Venus and Mars. He alleged (without even a morsel of proof) that Venus was once a comet engaged in an erratic orbit. As Venus approached Earth—and later when it displaced Mars from its trajectory—it caused all sorts of mayhem, seen to primitives as divine miracles.

The vast majority of scientists found Velikovsky's vision to be, in short, ridiculous. Comets didn't turn into planets, they argued, and planets couldn't behave in such a complex manner. Many of them were so upset by the mass popularity of Velikovsky's beliefs that they subsequently built mental barriers against any sort of catastrophic theory of extraterrestrial bombardment. Gradually, though, the scientific community became more open-minded and began to entertain notions such as that of the Alvarezes'.

The Alvarez hypothesis has been substantiated by many recent findings. Excess iridium has been measured in clay deposits along the K/T boundary in excavations all over the world. Shocked quartz, a signal of a violent event that shattered the quartz grains, has similarly been observed at the expected depths throughout the globe. A soot layer, evidence of rampant forest fires, has also been widely discovered.

Most important, geophysical research has found a likely candidate for the impact crater in a region centered on the village of Chicxulub in the Yucatan peninsula of Mexico. The basin of a massive collision has been detected there hundreds of feet below the ground. Because the crater's measured diameter is approximately 110 miles, scientists believe that it is the result of the crash of an asteroid more than six miles in diameter—close to the size hypothesized by the Alvarez team. Geological dating methods estimate that the body's impact occurred roughly sixty-five million years ago, about the same time as the K/T extinction.

Has the mystery of the disappearance of the dinosaurs been resolved? Many scientists now think so. The discovery of the Chicxulub crater and the evidence presented by K/T boundary deposits have convinced a large segment of the scientific community that the mass extinction at the end of the Cretaceous Period had a single primary cause of extraterrestrial origin.

In spite of the predictive success of the Alvarez hypothesis, several groups of dissenting scientists have sprung up in England, Japan, and the United States. Known as "gradualists," they are opposed to the hypothesis' catastrophic premise. One of these gradualists, Norman MacLeod of the Natural History Museum in London, argues that extinction was slow-going, taking place over millions of years, and stemming from a variety of causes. He feels that the impact of an object of extraterrestrial origin was probably only one of many factors, perhaps not even a decisive one. Put bluntly, "whatever wiped out the non-avian dinosaurs was a lot more complicated than a single hammer blow from an asteroid,"[1] he asserts.

Anthony Allen of the University of London and Shin Yabushita of Kyoto University in Japan have developed an alternate model for the saurians' demise, based on a different cosmic cause.[2] According to their extinction scenario, during the late Cretaceous period the Earth (along with the rest of the Solar System) passed through the dense core of a giant molecular cloud (GMC). This GMC was similar to the swirling collection of gases that evolved into the present day Sun and its orbiting planets.

In Allen and Yabushita's opinion, the GMC the Earth encountered eons ago was a massive, trillion-mile-wide concoction of molecular hydrogen mixed with dust. As our planet passed through the cloud, its atmosphere became choked with foreign chemicals. According to these scientists, to make matters worse, the Earth likely journeyed through the center of the whirlwind—the densest part, with the highest concentration of exotic material. Because of the size of the cloud, this exposure likely lasted for millions of years.

The effects on our poor planet of being dipped into this extraordinary chemical bath were quite horrendous, according to Allen and Yabushita. They were akin to the deadly impact on soldiers of racing through fields of poison gas. Exposure to such large quantities of molecular hydrogen served to rob Earth's atmosphere of fully one-third of its oxygen content. This substantial loss occurred because hydrogen and oxygen are a volatile combination, readily combining to form water. Earth's air, bombarded for hundreds of thousands of years by the cloud's material, soon became saturated with watery vapors.

One might wonder how such a simple, seemingly innocuous, reaction could have led to the mass extermination of numerous species. Water itself certainly isn't lethal. However, oxygen deprivation is. If enough oxygen turns into water, many forms of life cannot survive.

Allen and Yabushita conclude that the dinosaurs and other large creatures were short of breathable air and consequently asphyxiated. Smaller animals, needing less oxygen, managed to survive. They repopulated the world, unscathed by the experience, when the GMC pulled away. That is why the mammals, not the giant reptiles, rule the Earth today.

Allen and Yabushita cite rather tenuous evidence to justify their theory. They point to a drop in oxygen content in the air bubbles found in amber samples dating back to the late Cretaceous period. They say this reduction is consistent for all known specimens stemming from that period. However, these results are accepted by only a small segment of the scientific community. Many experts in the field think that these amber samples are not pure enough to form conclusions about the Earth's atmosphere during the time in which they were formed.[3] Therefore the Allen–Yabushita model attracts few adherents.

Another gradualist approach,[4] proposed by Bill Napier of Armaugh Observatory in Northern Ireland, has the advantage of not requiring such an elaborate mechanism. Rather, it supposes that cometary dust, released by a "supercomet" (an enormous comet) as it approached our part of the Solar System, served to clog the atmosphere, block out sunlight, cool the Earth, and eventually kill the dinosaurs. According to Napier, this dust was gradually released by the supercomet after it became trapped in the region near the Sun. Over time, the Sun's heat acted to sublimate the body, discharging its debris into space like the grime that might drip from the ceiling of a defrosting freezer. Eventually, as the Earth's atmosphere absorbed greater and greater quantities of this matter, the Sun's light found it harder and harder to break through. Soon, the air became frigid, and only the most adaptable of creatures (such as the furry mammals, but not the dinosaurs) could survive.

Acknowledging that there is no geological evidence that glaciers formed during the late Cretaceous period, Napier believes that the dusty shroud produced by the supercomet's sublimation was not thick enough to spawn a full-fledged ice age. Rather, he thinks that the debris served to cool off the Earth substantially—sufficient, over time, to obliterate the dinosaurs—but not enough to generate glaciers. Thus, in Napier's view the period in which the saurians died out was a "frost age" rather than an actual ice age.

According to Napier, once the supercomet lost most of its ice and exterior dust, it disintegrated. The power of its blast formed the Chicxulub impact crater. In contrast to the catastrophists, though, Napier believes the dinosaurs were already well on their way out when the Chicxulub object hit. Its impact, in his view, formed the fireworks at the end of an age of mass extinction, rather than that era's primary cause.

There are other theories. Deep in the heart of Southern India lie the Deccan Traps, a vast plateau formed of dried lava from an ancient series of volcanoes. For almost two decades, geologist Dewey McLean of the Virginia Polytechnic Institute has argued these plains hold the long-sought secret of why the dinosaurs and other creatures met their demise during the Cretaceous/Tertiary boundary period.[5] In yet another rebuttal to the Alvarez hypothesis, he asserts that a prolonged age of supervolcanism associated with that region led to the venting of mammoth clouds of sulfur dioxide, carbon dioxide, and other gases high up into the air. The contaminating effects of these materials led to massive environmental pollution, including noxious acid rain and a significant, prolonged increase in global temperatures. In particular, the build-up of carbon dioxide in Earth's atmosphere tended to block surface heat from escaping, resulting in pronounced terrestrial warming. As in the interior of a glassed-in hothouse, average temperatures steadily rose throughout Earth's surface. (This phenomenon, commonly known as the "greenhouse effect," will be discussed in Chapter Six.) In the midst of this havoc, McLean contends, dinosaur embryos and other forms of life simply could not survive. Thus a terrestrial event—supervolcanism—rather than an extraterrestrial event killed the great beasts.[6]

It might seem hard to believe that the action of a single volcanic region could cause such a global catastrophe. However, the Deccan Traps were fashioned by erupting volcanoes of unmatched power. According to geological theory, these eruptions occurred when an enormous fireball rose up from the center of the Earth, generating a formation known as a mantle plume. Like a bubble emerging from a vat of boiling water, this plume ascended from the Earth's core to its crust. As it rose, it tunneled out a conduit for the blazing energy of the core to escape to the surface. More than a quadrillion cubic feet of molten lava were spewed into the air, rapidly spreading over a region of the Earth's surface thousands of miles across. At its thickest, the scalding liquid constituted a veritable ocean, forming a crust miles deep as it dried. At no other time in recorded history has the inner Earth disgorged so much of its fiery material.

The noxious gases released in this colossal burst of volcanic activity likely had the aroma of billions of rotting eggs. These foul vapors, coupled with eye-stinging acid rain and intense heat from the resulting greenhouse effect meant the creatures walking the Earth at that time were probably pretty miserable.

Most geologists agree with the essence of the scenario just presented. McLean takes it a step further. Not only were Earth's creatures unhappy, he argues; a large number of them died out. Each living being had to fight for survival, with many not being up to the struggle. Dinosaurs in particular lacked the capacity to sweat off the searing heat. In a natural mechanism for trying to cool down, blood became diverted from the centers of their bodies to their skins. The diversion of blood in female dinosaurs meant that their uteri were deprived. Consequently, the heat-sensitive embryos within female dinosaur wombs failed to survive, and the great beasts slowly died off from attrition, according to McLean. In other words, as the dinosaur mothers struggled to cool themselves, they lacked the physical capacity to save their embryonic progeny as well.

McLean believes that the Alvarez hypothesis is flawed because it fails to account for the possibility that species extinction occurred through the long term effects of volcanic emissions, rather than the short term pandemonium due to the sudden blow of an extraterrestrial object. Several other scientists, such as Napier and MacLeod, concur with McLean's point that the impact theory is too simplistic, and would like to see other possibilities, such as supervolcanism in the Deccan Traps, investigated as well. They believe that it is far too early to close the books on any of the spectrum of possible causes of the mass extinctions at the end of the Cretaceous era.

In spite of these critiques, the Alvarez model continues to serve as the leading explanation of why the dinosaurs died out. It is not hard to see why. For the bulk of the geological community, the Chicxulub crater serves as the smoking gun. It provides a simple, visual remnant of a catastrophe that clearly devastated our planet. Even gradualists concede that the world must have suffered a dreadful fate of untold proportion when an asteroid or cometary intruder carved out that gigantic basin and sent myriads of dust particles into the air. As catastrophists point out, it could hardly have been pure coincidence that this event occurred during a time in which many species died out. Compared to the Alvarez explanation, other models seem far fetched. Thus in the minds of many scientists, the great extinction mystery has been solved.

Patterns of Destruction

The Cretaceous/Tertiary event was hardly the only case of mass extinction in geological history. Nor was it the greatest example of species annihilation. At the end of the Paleozoic era, 250 million years ago, life on Earth was nearly eradicated. More than 90 percent of all known species were wiped out. Among the more familiar species annihilated were the trilobites—three-lobed sea creatures with segmented bodies (related to crabs, lobsters, spiders and insects). Numerous trilobite fossils have been found dating back before this time; none has been discovered after that time.

Paleontologists have noted three other cases of mass extinction in which more than half of the known species died out. These occurred 215 million years ago at the close of the Triassic period, 360 million years ago at the close of the Devonian period, and 435 million years ago at the close of the Ordovician period. In other examples, dating back twelve million and thirty-eight million years from the present, somewhat fewer species became extinct.

Because of the success of the Alvarez hypothesis in explaining the K/T event, a number of scientists have advanced the proposition that *all* of the mass extinctions were caused by the bombardment of comets and/or asteroids. Geological investigation seems to bear this out. Excess iridium has been detected in a number of clay and rock layers, which represent times of large-scale eradications. There also seems to be at least some correlation between the number of impact craters found dating back to particular times, and the rate of species extinction during those intervals.

The likelihood that the great mass extinctions were triggered by extraterrestrial impacts has served as a catalyst for a number of scholarly conferences in the 1980s and 1990s that have considered the risks and consequences of possible future disasters. The first of these meetings, the NASA-sponsored workshop, "Collision of Asteroids and

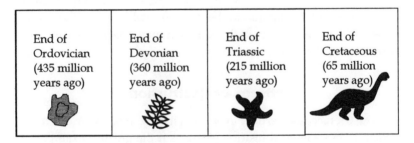

End of Ordovician (435 million years ago)	End of Devonian (360 million years ago)	End of Triassic (215 million years ago)	End of Cretaceous (65 million years ago)

Great species extinctions, in which large segments of Earth's plant and animal populations died out, have taken place regularly in the past.

Comets with the Earth: Physical and Human Consequences," took place in Snowmass, Colorado, in 1981, shortly after the Alvarez hypothesis was proposed. With astronomer Gene Shoemaker (later renowned for his comet discoveries) serving as general chair, sessions were headed by comet watchers, asteroid trackers, crater experts, and other Earth and space scientists. Each brought his own expertise to bear on the odds of catastrophic impact in years to come. The conference proceedings, never published, detailed the hazards of cosmic collisions, including the possibility of future extinction events, and urged that preventative measures be taken. These suggested precautions included the tracking of NEOs (Near Earth Objects), and the development of defense systems to divert such bodies before they do harm.

In 1992, upon request of the United States Congress, NASA asked a group of scientists to examine further the likelihood of Earth–comet and Earth–asteroid collisions, and to consider how these events might best be anticipated and, if possible, avoided. The Spaceguard Survey Working Group, chaired by David Morrison of the Ames Research Center, concluded that the long term peril of such impacts was significant. They pointed out that though in a given year, the chances of other natural hazards (earthquakes, for example) would be much greater than those of collisions, the rare impacts of extraterrestrial bodies would be so devastating that they would cause much more damage in the long run than that created by more common catastrophes. Because a single such crash would wreak extraordinary damage to our world and its inhabitants—possibly killing billions—Morrison and his group proposed the idea of a global network of telescopes especially designed to detect and track Earth-threatening extraterrestrial bodies.

Morrison, along with planetologist Clark Chapman, reported that the chances of civilization being destroyed by infalling extraterrestrial objects were approximately 1/300,000 per year.[7] In other words, once every three hundred thousand years, on average, a comet or asteroid large enough (estimated to be 1 kilometer in diameter or greater) to wreak enough havoc to wipe out advanced society (though not necessarily all of humankind) is due to strike Earth.

These chances, though small, are still frightening. Imagine the bulk of humankind's crowning achievements—scientific, artistic, political, and social institutions—crumbling into dust in the face of global cataclysm. The single hit of an asteroid even close to the size of the one that likely dropped on Chicxulub could propel trillions of tons of displaced material into the air. Uprooted soil and debris would shield the sun's direct rays—darkening the skies for a year, blocking the process of photosynthesis, and resulting in the wiping out of most common species of plant life. Global agriculture would vanish and widespread famine would set in.

It is hard to fathom the overwhelming sense of panic that would ensue from such a "year without sunlight." As food, drink, and medicine became scarce, hunger and disease would decimate our race. Widespread poverty would almost certainly lead to growing social unrest. As civil society crumbled, crime rates would likely soar. In the worst case, mobs would rule the streets unchallenged, bandits would roam the markets unhindered, and scoundrels would control the town halls unopposed. Millions of shops, houses, museums, libraries, and schools (that is, those that survived the blazes ignited by the fireball) would be looted and burned to the ground. Governments would fall, economies would crash, and general chaos would take hold. In short, civilization would likely never be the same again.

Given the manifest horrors of such a situation, scientists have been actively seeking ways of reducing the chances of such a catastrophe. Since the time of Morrison's study, NASA has spent about $1 million per year to survey and study nonplanetary bodies within the solar system. These funds are distributed to several research groups dedicated to the telescopic tracking and analysis of asteroids and comets. Current budgetary restraints make it unlikely that NASA will increase its funding of the surveys.

Included in those grants has been the Spacewatch project, a mission conducted by astronomers Tom Gehrels, Robert McMillan, and their group at the University of Arizona's Lunar and Planetary Laboratory and Steward Observatory. The Spacewatch team has pioneered the use of CCD (charge-coupled detectors) to collect, image and analyze the faint reflected light of comets and asteroids. Conducting their sky scans with the 0.9-meter (three-foot-diameter) Spacewatch telescope at the Steward Observatory on Kitt Peak, they

A view of the two Spacewatch telescopes on Kitt Peak in Arizona. The 1.8-meter telescope is in the foreground, and the 0.9-meter is behind it. (Courtesy of Tom Gehrels.)

have made a number of remarkable discoveries so far, including the closest approach of any asteroid to the Earth, the faintest known comet at the time of its detection, and the smallest known asteroid (about twenty feet in diameter). Luckily, the group has yet to observe a celestial body fated to collide with our planet.

In spite of our good fortune so far, we must not be complacent about the possibility of collisions. The sighting of an asteroid headed on a fatal course toward Earth could occur at any time. Whether such a dire message arrives months, years, or centuries from now, we must remain ever vigilant.

In March 1998, a headline-making announcement by Brian Marsden of the Minor Planet Center at the Harvard–Smithsonian Center for Astrophysics brought a shocking sense of urgency to this issue. Marsden reported that a mile-wide asteroid called 1997 XF11, discovered by Jim Scotti of Spacewatch in December 1997, was fated to pass within thirty thousand miles of Earth on October 26, 2028. The asteroid's predicted approach seemed so close—far fewer miles than the Moon's orbital distance—that collison presented an all too frightening possibility.

The sense of crisis lasted only momentarily, however. Within days, Marsden announced revised projections that the asteroid will miss Earth by a comfortable six hundred thousand miles. The chances of impact, he cheerily reported, were much smaller than first thought. Still, he warned, any small perturbation in the asteroid's trajectory could redirect it to a doomsday course. Even if that object ultimately presents no danger, he advised, someday we may not be so fortunate.

If a deadly object were predicted to impact upon Earth, means for diverting it would have to be put into place. The magnitude of such a threat would mandate the rapid deployment of a space mission designed to deflect the comet or asteroid as much as needed to save Earth. In such a system, weapons would be launched into orbit, programmed to deter an extraterrestrial intruder if it ever came close enough.

In Gehrel's influential book, *Hazards Due to Comets and Asteroids*, he, Morrison, Marsden, and dozens of other prominent researchers, including Clark Chapman of the Planetary Science Institute in Tucson, Arizona, Carolyn and Gene Shoemaker (codiscoverers of Comet Shoemaker–Levy), and Vadim Simonenko (current head of the Russian Space Shield Foundation), detail methods for detecting asteroids and comets, address the dangers of collisions, and offer a number of interesting suggestions for deflecting threatening objects. Options discussed include pulsed (high energy) laser beams, nuclear explosives, or projectiles designed to deliver enough power to blast the comet or asteroid off its doomsday course.[8, 9, 10]

The consensus of these authors is that a stand-off (taking place in space, rather than buried within the object) chemical or nuclear explosion would be the most effective means of diversion.[11] By launching a device into space and exploding it very close to the comet or asteroid, sufficient energy could be applied to change the intruder's course. The rogue object would then proceed harmlessly off into deep space.

Let's hope, then, that if Spacewatch or another group ever detects a body fated to collide with Earth, an effective defense system will be well established. The alternative—the devastation of our species in the same manner that the great thunder lizards were vanquished eons ago—is far too dreadful to contemplate.

Prospects for Disaster

Sometimes it is hard to accept the inevitable. A terminally ill patient often fights for his or her last breath, no matter how much the pain. And rightly so. It is life's mission to perpetuate itself, and, if possible, to propagate itself. Living organisms are simply not made to die without a struggle. A computer program can terminate itself, but not a psychologically healthy human being. As Thomas' poem implores, we "do not go gentle into that good night."[12]

What if we spend billions of dollars to defend the Earth against the possible impact of comets and asteroids, and then an unexpected extraterrestrial body somehow manages to evade these systems and break through? If the intruder was large enough and happened to land in the ocean, massive tidal waves would wreak havoc upon thousands of coastal and island communities. If the body hit the ground, it would churn up enormous amounts of soil. The atmosphere would be choked with dust, blocking sunlight for many months. Many forms of life on our planet would be decimated. Conceivably, humankind would be destroyed.

Or what if the coming millennia represent the dawn of another ice age, which transpires as part of a natural climatic cycle? Perhaps ice sheets would cover much of the northern hemisphere, and the human race would lack enough arable land to support itself in its present state.

Other disasters could similarly threaten our civilization. For example, the ozone layer, the thin segment of the upper atmosphere that protects us from the Sun's most harmful rays, has been under siege for decades by highly reactive human-made chemicals (such as artificial refrigerants). Though these ozone-deleting agents have been banned in the industrial world, the gases already released are extremely long-lasting. Scientists project that they will linger for years and wreak additional destruction. If Earth's upper-atmospheric ozone continues to diminish sharply under these effects, cancer and cataract rates would skyrocket beyond belief. Damaged by harsh ultraviolet light, world crops would become irradiated beyond repair, resulting in catastrophic food shortages. Radiation-induced mutations would also lead to a rise in animals (including humans) born deformed. In short, the surface of our planet would become a hostile place to live.

Additional forms of environmental pollution could create havoc of an equally lethal sort. For many years, the exhausts from factories and cars have been accumulating in the atmosphere, trapping the Earth's heat and creating what is called the "greenhouse effect." Global temperatures have been rising for decades. If they increase enough, the polar ice cap could melt significantly—drastically raising ocean levels and flooding thousands of population centers—especially low-lying island and coastal communities.

Nuclear bombs and chemical weapons currently possess the capability of wiping out sentient life on this planet. If enough were acquired by a dictator or band of terrorists, the last days of civilization could be at hand. A fatal illness for which there is no cure could similarly destroy our race. And ultimately the Sun will die, robbing the Solar System of its primary source of heat and light. Before it has completely expired it will swell up into what is called a red giant star, possibly gobbling up Earth in the process. Over the eons, our planet is surely doomed; it's only a question of how and when.

The valley of apocalypse is broad and deep, and there are countless paths that lead right down into it. We might lose our footing and stumble, hastening our plunge. The end of the road could come tomorrow through such a misstep. Or we might take the slow, winding route down, approaching the bottom in millions or even billions of years. Either way, though, the human journey will someday reach its limit.

Yet, in spite of its apparent inevitability, we rebel against this dreadful fate with all of the strength we can muster. Throughout the ages, we have prayed to various spirits, hoping that human kindness and sincerity will triumph over the forces of natural destruction and cosmic indifference (to our plight). Most ancient narratives of the "end of days" have an escape clause—a last minute reprieve from the jaws of death. We must trust in a force—these stories admonish us—that will rescue our race before it's too late. Our species will be ravaged, but, like the legendary Phoenix, will rise again.

Perhaps there is no better symbol of apocalypse averted than the coming and passing of a near-Earth comet. Imagine the sense of dread experienced by primitive peoples when the blaze of a fireball first appeared in the skies. As the comet's visage waxed in the heavens, a sign that it was approaching, their feelings of terror must have grown to a crescendo of panic. Surely doomsday must have seemed at hand. Finally, imagine their enormous relief when the fireball left the skies and the world was found to be completely unharmed.

Even in the modern era, the coming of comets has triggered panic. In 1997, the spectacular appearance of Comet Hale–Bopp over the skies of the Northern Hemisphere inspired one of the worst cases of mass suicide in history. Believing that an alien spaceship lurked behind Hale–Bopp, thirty-nine members of the Heaven's Gate group chose to end their own lives abruptly. Led by charismatic UFO fanatic Marshall Applewhite, they hoped to hitch a spiritual ride on the extraterrestrial craft as it left our system.

Such an irrational reaction to a comet harkens back to the way early peoples reacted to similar astronomical phenomena. Imagery from ancient times provides us with a record of the hopes and fears of early societies. Comets were not the only celestial objects that were viewed as portents. The motions of the stars and planets through the heavenly dome were scrutinized carefully for signs of important omens. Natural disasters on Earth, such as floods and famines, were seen as being influenced by occurrences in the celestial sphere. Furthermore, by appealing to heavenly forces, primitive peoples believed they could possibly ward off potential catastrophes in nature. This complex, dynamic relationship between the human, natural (terrestrial), and heavenly realms is a fascinating hallmark of early civilization. Omnipresent in ancient art, literature, science and other forms of culture, it portrays the symbiotic interplay between earthly and cosmic order.

PART I

WRATHFUL GODS

RELIGIOUS APOCALYPSE

CHAPTER ONE

THE FLOODS OF BABYLON

EARLY APOCALYPTIC IDEAS

> *On that day all the fountains of the great deep burst forth,*
> *and the windows of the heavens were opened. And rain fell*
> *upon the earth forty days and forty nights . . . And the*
> *waters prevailed so mightily upon the earth that all the*
> *high mountains under the whole heaven were covered . . .*
> *And all flesh died that moved upon the earth, birds, cattle,*
> *beasts, all swarming creatures that swarm upon the earth;*
> *everything on the dry land in whose nostrils was the*
> *breath of life died.*

—Biblical Account of the Flood (GENESIS 7: 11–22)

After the Deluge

Spring rains often seem to go on forever. In many temperate parts of the Northern Hemisphere, when winter's icy shield is finally shattered by the mild days of April and May, the sky opens up and lets loose the frigid reservoir it has been storing for months. Interminable days of drizzly dullness are often followed by ceaseless nights of sustained downpour. Torrent piles upon torrent, transforming thawing snow into a thick gray slush. And gradually this watery glop melts away, dissolving into vast muddy puddles and dirty tepid streams. These rivulets merge with the rainfall water, forming dreary currents that steadily snake their way through streets and sidewalks, fields, and forests.

Sometimes if it rains long and hard enough, and the snow melts quickly enough, proper drainage proves impossible. That's why spring thaws are frequently followed by spring floods. Occasionally these floods grow so powerful they overwhelm their surroundings. Melting glaciers pour their watery cargo down the sides of mountaintops, inundating small villages that lie in their paths. Riverbanks overflow, spilling onto sidewalks and streets, knocking over walls and fences, and saturating the foundations of houses, churches, and schools. Topsoil and rubbish mingle as they drift along twisted currents, flowing through sodden gullies carved by the devastating force of water.

Even in global regions free of snow and ice, there is often a rainy season. During particular times of the year in these places—usually winter—it just pours and pours and pours. In many parts of the world torrential rains flood certain communities on a regular basis. People living there must learn how to cope with the ravaging effects of rapidly flowing water, and hope that the rainy period is mercifully brief.

It is no wonder that in ancient times, the idea of a great deluge constituted such an evocative symbol of widespread destruction. Numerous mythological accounts from before the time of Christ— most famously, the biblical tale of Noah—refer to great floods which wiped out entire civilizations. The ancients feared this would happen again. They thought if the human race were annihilated, it would likely have a watery grave.

Universal deluge was feared because floods represented a relatively common type of disaster. Many early peoples, familiar with the ravaging impact of water, could readily picture the horrors of a downpour that refused to let up for weeks. They could quite vividly imagine the nightmare of buckets and buckets of rain pouring down upon the Earth relentlessly, flooding streams and rivers, valleys and plains. Considering the damage that an ordinary deluge could do, they dreaded the destructive force of a universal flood. The ruination of agriculture, the inundation of towns and cities, and ultimately the near-annihilation of the human race were all part of their diluvial or great flood imagery.

Legend of Atlantis

Aside from the tale of Noah, perhaps the most famous flood story in antiquity is the legend of Atlantis. The Atlantis chronicles first appeared in Plato's *Timaeus*. Plato spoke of a vast island, "larger than Libya and Asia together" (meaning northeast Africa and Asia Minor, respectively), that flourished more than eleven thousand years ago, and later sunk beneath the Atlantic Ocean.

Scholars are divided as to whether or not Plato's account refers to an actual place. Even the ancient Greeks debated its veracity. Aristotle, a student of Plato, considered the tale to be purely allegorical. On the other hand, Crantor, a scholar who edited *Timaeus*, claimed that it was the absolute truth.

In the centuries since the death of Plato (347 B.C.), thousands of commentaries have been written about the Atlantis legend. Numerous places around the world have been proposed as the true setting for the tale—everywhere from North America to ancient Troy. Some writers have alleged there really was a vast island in the middle of the Atlantic Ocean, from which European, African, Middle Eastern, and Native American cultures all stem. In *Atlantis: The Antediluvian World,* a highly influential work published in 1882, American novelist Ignatius Donelly described Atlantis as a highly advanced civilization where early humankind flourished in a veritable paradise. However, years of diving expeditions and other maritime searches have yielded no evidence of such a land's existence.

Most of Plato's histories are known for their impeccable accuracy. Therefore it is surprising to scholars that he would write a fictional treatise and assert it to be the truth. Nevertheless, based upon all available evidence, we must consider his account to be a mere fable.

Regardless of its veracity, Plato's tale is an outstanding epic of an exalted civilization faced with extinction by deluge. In this manner, it stands as a warning to advanced societies of the possibility of sudden destruction. With the example of Atlantis in mind, no culture, no matter how successful, should feel absolutely secure in its prominence.

According to Plato, Atlantis was destroyed in the height of its power. Before its devastation, he wrote, it ruled large parts of Europe and Africa. Its martial arts were exemplary, and its warriors powerful. Supposedly, its remarkable leaders were literally descended from the gods.

Because of the island society's considerable might, it aspired to rule the world from its base in the Atlantic. However, its plans were thwarted by sudden catastrophe. Zeus, the Greek "god of gods," decided that the Atlanteans were becoming too haughty and self-righteous, and wanted to teach them a lesson. He sent earthquake shocks and floods, until the island nation became submerged.

Plato's account is quite vivid in its description of the end of Atlantis:

> There occurred portentous earthquakes and floods, and one grievous day and night befell them, when the whole body of [Greek] warriors was swallowed up by the earth, and the island of Atlantis in like manner was swallowed up by the sea and vanished; wherefore also the ocean at that spot has now become impassable and unsearchable, being blocked up by the shoal mud which the island created as it settled down.[1]

Plato makes it clear that the death of Atlantean culture was far from a solitary event. Rather, he argues that the human race has been challenged time and time again by natural disasters such as floods and plagues. The Atlantis episode, he suggests, is just one link in a long chain of golden ages and cataclysms. After each success for humankind, it is only a matter of time, he feels, before the next calamity will strike.

Gilgamesh and Noah's Ark

A common feature of flood myths is the idea of a society being punished by God (or the gods) for its excesses. This is certainly true in the case of the Atlantis story. The Atlanteans were struck down by Zeus for what we would call an "attitude problem."

In the biblical tale of Noah and his ark, God finds the peoples of the Earth to be corrupt and violent. He decides to wipe out all of humankind, sparing only Noah—a just man—and his family. Noah is instructed to build a boat, in which his household, and representatives of all types of animals, are to take shelter from a global deluge. After Noah and his family retreat to their ark, a flood ensues for forty days and forty nights. The earth is covered with water until the flood subsides. Finally, the world becomes dryer and Noah and his family can leave the ark. God asks Noah to repopulate the earth, and to obey his laws. In return he promises Noah and his heirs never to destroy the world by flood again.

In the chronicle of Noah, once again we see the theme of God destroying a civilization in order to punish its shortcomings. In this case, the society being castigated is the human race itself, save a few of its most righteous members. Another important element of this tale, found in many flood myths (including one Greek legend, called the tale of Deucalion and Pyrrhain, in which Zeus swamped the world with rain, destroying everyone except for two righteous survivors), is the idea of a solitary hero and his mate who build a boat, escape the deluge, and repopulate the earth. For this reason the story of Noah represents something less than total annihilation; it is apocalypse with an escape clause.

Furthermore, in the Noah saga, God pledges that global flooding will never happen again. Clearly this message is meant to offer solace to those reading the tale. Interestingly, though, God doesn't promise never to destroy the world again, just never to do so again *by deluge.* Thus, we find additional apocalyptic scenarios—though never by water—later on in the Bible.

Some biblical scholars consider the source of the Noah story to be the ancient Babylonian Epic of Gilgamesh. (The Babylonian kingdom once comprised what is now known as southern Iraq.) This tale is the oldest known flood chronicle to be preserved in writing. It was scribed in cuneiform almost four thousand years ago on a series of clay tablets. These tablets were unearthed in the 1840s by the young English archeologist Austen Layard. Layard was excavating the ruins of Ninevah (the ancient capital of Assyria, and later an important center of the Babylonian kingdom) when he stumbled upon this priceless collection of epic poetry. In the early 1870s, British archeologist George Smith translated the tablets, and came to the realization that they represented a story of the great deluge even older than the biblical version. He presented his findings in 1872 to an astonished audience at a meeting of the Society for Biblical Archaeology.

The part of the Epic of Gilgamesh that deals with the great flood begins with the namesake of the tale engaged in a long, perilous trek. Gilgamesh, a warrior king whose best friend and traveling companion Enkidu has just died, has set out on a quest for the secret of eternal life. He seeks the dwelling place of the only man who has ever achieved immortality, Utnapishtim, survivor of the great flood. Utnapishtim lives with his wife in the "place beyond the Waters of Death." Once Gilgamesh finds him, he listens to the immortal's account of the deluge.

Utnapishtim begins his report by explaining that the decision to flood the earth and rid it of its human population was made by the gods sometime in the distant past. He doesn't relate why they made their momentous decision; other flood myths from the region explain that the gods often tire of human chatter. Once the resolution was confirmed, the god of the waters, Ea, appeared to Utnapishtim in a dream and warned him of what was to come. He was told to tear down his house of reeds, and use the materials to construct an ark. Curiously, the ark was to be built in the shape of a cube, of dimensions specified by Ea. He was further instructed to load the boat with his family, his personal craftsmen, and representative animals.

Utnapishtim then relates how he obeyed Ea's instructions to the letter, and boarded the ark as soon as it began to rain. The rains came hard and fast, and soon covered the earth with water. After seven days of deluge, it finally stopped raining and Utnapishtim's ship came to rest on the mountain of Nisir. He waited another seven days and then disembarked.

As the floodwaters receded, Utnapishtim and his family proceeded to repopulate the Earth. Meanwhile, the gods decided—convinced by Ea—never to destroy the world again by deluge. Ea counseled them to use other methods of annihilation in the future.

His suggestions included death by plague, mutilation by wild animals, and starvation through famine. The gods consequently agreed to these "progressive" alternative approaches, and bestowed on Utnapishtim, for all of his troubles, the secret of everlasting existence. After Utnapishtim concludes his tale, he instructs Gilgamesh on where to find the plant of eternal life. Gilgamesh, eager for immortality, eventually hunts down and finds the plant. But then, before he has a chance to taste the plant, a serpent appears and snatches it from him. Gilgamesh comes to realize that death is the lot of all men, save a lucky few exceptions gifted by the gods.

Flood Fossils

There are hundreds of other flood myths, associated with numerous peoples throughout the globe. In almost all of these legends, humankind has been punished for its transgressions, and is fated for watery extinction. Then, in the majority of these epics, a solitary man or family becomes exempted from the apocalypse, and is chosen to repopulate the Earth. Philip Freund, in his book *Myths of Creation*, describes a few of the legendary methods of salvation from universal deluge:

> Nichant, the hero of the Gros Ventres, swims while holding onto a buffalo horn. Rock, the bold ancestor of the Arapaho Indians, fashions himself a boat of fungi and spider webs. The lone progenitor of the Annamese saves himself in a tom-tom. The Hero of the Ahoms in Burma uses a gigantic gourd which, by magical intervention, providentially grows out of a little seed. Trow, of the Tringus Dyaks of Borneo, is tossed on the waters in a trough as is the heroine of the Toradjas of Celebes, though hers is—most unromantically—a swill trough. The ancestors of the Chané of Bolivia find refuge in an earthenware pot that floats. . . .[2]

The near-universality of flood mythology has caused some geologists to speculate that widespread deluge, during particular periods, inspired these legends. Perhaps the experience of the melting of glaciers at the end of the last ice age, and subsequent rising of the seas, was incorporated somehow into the oral tradition of primordial man. Or more likely, local floods, from time to time, caused such widespread tragedy that they were viewed as universal. Word of mouth accounts may have transformed regional disasters into events of epic proportions. For instance, when travelers from different nations swapped stories again and again, their recounting of floods may have taken on greater and greater terror.

Those who believe that a great deluge actually took place often point to fossil evidence of maritime life displaced where it shouldn't be. These "diluvianists" point to strange cases of uprooting such as sea shell fossils discovered in French inland rocks and whale skeletons found in the Sahara desert. But "gradualists" of the school of Charles Lyell, the noted 19th century naturalist, rebut that the world's waterways and land formations have slowly evolved over the eons. Given the gradually changing face of the earth's surface— due to the shifting of tectonic plates and to natural erosion— it is no wonder, they argue, that maritime fossils have been found in land-locked areas.

In spite of the strong influence of Lyellian thought on modern geology, attempts to prove scientifically the biblical flood account persist. Some speculative thinkers have pondered the notion that a comet collision thousands of years ago caused the great deluge. This tradition dates at least as far back as a treatise, entitled *A New Theory of the Earth, from its Original to the Consummation of All Things*, published in 1696 by the English mathematician William Whiston. In this text, Whiston proposes that the flood was caused by rain pouring down from a nearby comet's vaporous tail. Though few of Whiston's contemporaries could see any scientific merit to his thesis, his supporters included such notables as Isaac Newton and John Locke. Acclaimed comet watcher Edmund Halley, writing around the same time as Whiston, similarly proposed that the deluge was set off by cometary bombardment.

More recently, the idea that a great flood was caused by extraterrestrial influences has become associated with the theories of the late fringe scientist Immanuel Velikovsky. (Recall that Velikovsky argued that Venus was once a comet, and that it passed very close to

Earth several times.) Velikovsky proposed that the close passage of a comet caused catastrophic tidal forces to devastate Earth. As the comet neared, ocean waters supposedly washed up onto the shores and flooded large land areas. These tidal waves were monumental, beyond the scope of any seen before. The biblical tale of Noah, Velikovsky maintained, is a loose narrative of these cataclysmic events.

Could the impact or near-collision of a celestial body with the Earth have caused a great flood? Highly unlikely. First of all, there are no indications in the geological record of a deluge of such universal proportions. Surely evidence of such a catastrophic event (if it had occurred) would be omnipresent in the rocky strata examined by geologists, much more so than the isolated examples of uprooting that they have observed. And further, if a comet or asteroid caused the flood then there would be global connections in the rock record between collision debris and residue from a deluge.

Also, scientists, such as D. E. Gault and C. P. Sonett of NASA's Ames Research Center, have run laboratory simulations, testing what the impact of a massive projectile on the Earth's oceans would be like. They have concluded that "superwaves"—enormous ocean formations much higher than the tidal waves typically generated by earthquakes—would be created by the blast. However, the waves produced, though significant, would hardly be enough to flood the whole world. Most likely, while coastal areas would be greatly affected, hilly and mountainous regions, as well as places far from the ocean, would barely be touched. Thus, it seems clear that extraterrestrial bombardment could not have caused a flood of biblical proportions.

Rhythms of Renewal

In the absence of a literal scientific explanation for the prevalence of flood legends, it is instructive to look at metaphorical reasons instead. By examining what the notion of deluge meant to ancient cultures, we may gain insight into the reasons why they believed a great flood had taken place.

Water possesses a chaotic power. It is a bearer of life and a bringer of death. It soothes the parched lips of desert nomads, and lures lost sailors to their early graves. It is a source of sustenance and a receptacle of waste. Trickling slowly through tree-lined canals or over stony waterfalls, it provides endless beauty. But whirling around a ship in a maelstrom, it offers matchless terror.

We are born in water, emanating from a sac of amniotic fluid. Human life emerges fresh and wet from the womb. In this manner, water is a creator of potentialities, a symbol of birth and rebirth.

No wonder the ancients considered deluge to be the chosen method of the gods for worldly renewal. When the gods wanted to recreate the earth, they bathed it in a birth-sac of fluid, and delivered a new breed of human. For the ancients, then, the great flood constituted a universal rebirth, with the gods as midwife.

Most early cultures perceived time as a cycle. They believed that the world has been created, destroyed, and recreated an infinite number of times. Thus, they envisioned that great floods occur periodically—renewing the Earth again and again after fixed intervals of time—repeating themselves for all eternity.

In general, these cycles of demolition and rebuilding follow a standard blueprint from culture to culture. At the beginning of each cycle, the world (or part of it) is destroyed, except for a select few people. After the episode of destruction is over, a new civilization emerges from the ruins of the old, led by the survivors of the previous society. The new society grows and develops until it reaches its pinnacle. This "golden age" of peace and prosperity lasts for many centuries, perhaps even several millennia. Following the golden age comes a period of decadence, in which corruption and other forms of sin run rampant. This degradation leads to the society's ruin. Finally, according to the standard blueprint, civilization becomes wiped out again by a major catastrophe and the cycle begins again.

Religious historian Mircea Eliade documents in his book, *The Myth of the Eternal Return*, how notions of great cycles were omnipresent among ancient cultures. The idea that the world will renew itself someday provided early societies with a measure of optimism. No matter how bad things were, Eliade reports, these groups clung to the hope that a new cycle of time would usher in a new golden age. A flood, or some other type of disaster, would wipe out all evil, and thereafter herald in a time of goodness and joy.

In primitive mythology, floods were considered to be just one of a class of possible agents of global renewal. Other forces of destruction and recreation common to ancient myths include great fires, widespread famine, droughts, plagues and devastating warfare. In many schemes, cycles of time begin with one of these agents (deluge, for instance), and end with another (such as fire).

Consider, for example, the cyclical concept of history believed in by the peoples of ancient Babylonia. According to Babylonian mythology, worldly events repeat themselves in multi-millennial intervals known as "Great Years." Great years represent the divine equivalent of ordinary years for mortals. Rather than being measured by the time it takes for the Earth to travel around the Sun, they comprise instead the period in which the planets (those known at that time) repeat their relative motions in the sky. In other words, one Great Year is the interval between two complete alignments of all of the planets.

Babylonian mythology distinguishes between times in which the planetary alignment takes place in the part of the zodiac known as Cancer, and times when it occurs in Capricorn. The former is referred to as the Great Winter Solstice, and the latter as the Great Summer Solstice. Concurrent with the Great Winter Solstice is a period of deluge, when the world is covered by water. The Epic of Gilgamesh records one such instance of this flooding. In contrast, when the Great Summer Solstice makes its presence known in the heavens, the world becomes destroyed by fire.

Our perspective here does not permit a complete accounting of all of the cyclical time schemes of ancient cultures from around the

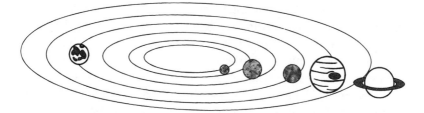

In many early cultures, including the Greek and Babylonian, planetary alignments were seen as heralding the beginnings of new ages.

world.[3] Let it suffice to say that most bear striking resemblance to the Great Year scenario of Babylonian culture. The main differences lie in details such as how long each cycle lasts, as well as the nature of the catastrophe that closes each cycle.

For example, the Hindu time scheme includes world cycles, known as *Kalpas*, each lasting 4,320,000,000 years. At the end of each *Kalpa*, the Earth bursts into flames, and a new creation emerges from the ashes of the old. Other cultures' timetables are distinguished by much shorter intervals; the world cycle of the Aztecs and Mayans is only fifty-two years.

Thus Spake Zarathustra

When we imagine global apocalypse, we picture the end of life on Earth, or at least the close of civilization. We consider the effects of "doomsday" to be resolute and eternal. Yet for most of the world's ancient peoples, global cataclysm was believed to be followed by epochs of universal renewal, as we saw with stories of a great flood.

The contemporary Western linear notion of history—comprising a unique progression from a fixed moment of creation in the past to a definitive ending in the future—offers a far bleaker version of apocalypse. In linear historical conceptions of time, doomsday represents the end of civilization (at least as we know it), rather than the close of one cycle and the start of the next.

The linear way of thinking about the end of the world has crept into the human psyche rather gradually. Cyclical time schemes were very attractive to early agricultural societies mainly because of the natural analogy between great cycles and the seasons. Primitive cultures felt comfortable thinking of history as seasonal—with the hot blasts of summer succeeding the flowery golden age of spring, and the cold rains (or snows) of winter following the solemn decline of autumn. It was only by shattering this metaphor that the modern idea of history as an indelible succession of events could begin.

One of the first religions to break the mold of belief in cyclic eternity was the Persian faith, Zoroastrianism. The Zoroastrian creed is based on the principles of Zarathustra, also known as Zoroaster, who lived (according to the conjectures of historians) sometime between

1700 and 1500 B.C. It also encompasses many rituals and beliefs from primitive cultures that lived in and around Iran thousands of years ago. A highly ethical religion, Zoroastrianism makes strong distinctions between the principles of good and evil. Its holy writings, the *Avesta*, serve as guidelines to promote lives devoted to justice and decency toward others.

The most important concept in Zoroastrianism is called *asha*. This word has no exact equivalent in English. Loosely translated, it means "the way things ought to be." It represents a way of living in harmony with natural order. Sincerity, personal integrity, truthfulness, and honor are all components of such a lifestyle.

Zoroastrianism makes strong distinctions between moral antitheses. It teaches the belief that people have the capability of doing either good or evil, and must learn to choose the former. They must aspire to follow *asha*, and eschew its opposite, *druj* (falseness). The righteous, who lead lives of *asha*, are called *ashavan*. On the other hand, those who practice evil, and bring disorder to the world, are referred to as *drujvan*.

The concepts of *asha* and *druj* pertain to universal laws, as well as to human affairs. *Asha* organizes the cosmos. It steers the planets along their paths, and makes the sun come up each day. *Druj* creates chaos, disharmony, and death. It is responsible for all that goes wrong with nature. *Asha* is a much stronger force than *druj;* that is why nature usually appears beautiful and harmonious.

One crucial difference between Zoroastrianism and religions based on time cycles (such as Hinduism) lies in what the former has to say about the end of the world. Unlike devotees of cyclical time-based faiths, Zoroastrians believe in a "last judgement," in which good is rewarded, evil is punished, and history draws to a close. The principles of Zarasthustra do not allow for evil to return again after the time of judgement; therefore, from that point on, new cycles of existence would be impossible. Once the force of *asha* reigns triumphant, the human drama is over.

Like the Christian book of Revelation, the Zoroastrian concept of apocalypse is rich with vivid imagery. In the closing days, the bodies of all who have ever walked on Earth will reassemble from dust and rise again. This grand resurrection will transform all of the dead back into fully functional living beings—and enable them to face judgment for their worldly deeds. Once the whole of humankind has been revived, it will be gathered together to witness the ultimate triumph of good over evil.

The rendering of the final verdict will be swift and decisive. The process will begin when all of the mountains of the world catch on fire, and their metal is melted. A flood of molten metal will then flow down from the mountains and cover the entire Earth. All those who have been resurrected will be required to pass through this vast molten stream.

For the wicked *drujvan*, this ritual will truly represent their last ordeal. They will step into the hot liquid and become painfully consumed. The presence of evil will literally become dissolved from the face of the planet forever.

The saintly *ashavan*, in contrast, will not mind plunging into the river one bit. For them, it will seem to be made of warm milk rather than of fiery metal. They will pass through the liquid completely unharmed.

After all of the wicked have perished, leaving only the righteous, the river of molten metal will pour into the bowels of the earth. The earth's topography will have changed by then. Because all of the mountains will have melted down, the earth's surface will consist of a flat plain. The landscape will become transformed as well. Stately trees and colorful flowers, beautiful beyond compare, will sprout up from the flattened surface. The world will seem a paradise, comparable only to heaven and the legendary Garden of Eden. The *ashavan*, having passed the test of fire, will be granted the gift of immortality. They will live in the earthly paradise forever, enjoying lives of harmonious bliss. Here we see in Zoroastrian belief an early formulation of the concepts of heaven and hell. Zoroastrians hold individuals fully accountable for their life's work. Those who spend their lifetimes in just pursuits are amply rewarded; they are blessed with eternal happiness. On the other hand, those who squander their mortal existence are ultimately condemned to the fiery lake of torment. Unlike the case of cyclical religions, such as Hinduism, there are no second chances.

Daniel's Dreams

Religious scholars believe that Zoroastrian thought about heaven, hell, and the end of the world had a strong influence on the early development of the great monotheistic faiths of Judaism, Islam, and Christianity. In the case of Judaism, it is clear that there were many opportunities for cultural crossbreeding. For a time, the ancient Hebrews were ruled by Persians, and were fully exposed to their faith. Moreover, years after the temple in Jerusalem was destroyed by Babylonian invaders, the Persian king Cyrus intervened on behalf of the Jews and allowed them to rebuild their temple. This generous act sparked sympathetic feelings among the Jewish people toward the Zoroastrians. Undoubtedly this encouraged the Jews to learn even more about the Persian faith.

Perhaps it is no coincidence, then, that the first lengthy discussions in the Hebrew Bible about the end of days occurs in the book of Daniel, dating back to a period when the Persian empire was formidable. The book of Daniel embodies the Old Testament's primary source of apocalyptic writings, containing several accounts of prophetic dreams and other premonitions about doomsday. Interestingly, there are many parallels between the apocalyptic ideas expressed in Daniel and those taught as part of the Zoroastrian creed.

Daniel, the protagonist of the story, is one of the Bible's most unusual heroes. When he was a boy, Jerusalem was sacked by Nebuchadnezzar, king of Babylon, who ordered its richest treasures and the most promising of its youths to be brought to the royal court. As one of this select group, Daniel grew up as a page in training, schooled by royal instructors. He soon discovered that he had a remarkable gift for prophecy and began to exercise his talents.

The Bible recounts a specular series of apocalyptic visions described by the youth. The most famous of these portents concerns four great beasts:

> Four great beasts came up out of the sea, different from one another. The first was like a lion and had eagle's wings . . . a second one [was] like a bear. . . . After this I looked, and

lo, another, like a leopard, with four wings of a bird on its back . . . After this I saw in the night visions, and behold, a fourth beast, terrible and dreadful and exceedingly strong; and it had great iron teeth and . . . ten horns.[4]

Daniel interprets his dream to be an omen of great warfare in the ages ahead. The four beasts, he relates, represent four kingdoms that will become embroiled in conflict. The fourth monster—the most terrible—stands for the powers that will emerge as victor. Thus Daniel prophesies that the last days of the world will constitute an age of bitter turmoil.

Daniel goes on to speak about a great resurrection of all who have ever lived that will take place in the closing days of the world:

Many of those who sleep in the dust of the earth shall awake, some to everlasting life, and some to shame and everlasting contempt. And those who are wise shall shine like the brightness of the firmament; and those who turn many to righteousness, like the stars for ever and ever.[5]

Daniel's description bears remarkable similarity to the Zoroastrian vision of the apocalypse. His account of the separation of good from evil parallels the Zoroastrian belief in the division of humankind into *ashavan*, who become immortals, and *drujvan*, who perish in shame.

Judaism has continued the tradition of faith in the coming of the "end times," as expressed in Daniel, and has passed on that conviction to Islam and Christianity. Over the ages, Jewish eschatological notions have come to encompass belief in a Messiah—a future leader designated by God to rule the world in its final days. The Jews have continued to wait for the arrival of such a figure, eschewing the Christian belief that this role has already been filled by Jesus Christ.

Throughout history, the Jewish people have encountered a number of charismatic figures who have proclaimed themselves "Messiah." The most famous of these was a young 17th century Jew from Smyrna (now part of Turkey), possessed with incredible personal magnetism, named Sabbatai Zevi. In 1648, Zevi announced that he had been chosen by God to lead the world in its end days. After persuading thousands of devotees to give up their possessions and join him, he was threatened with death by Turkish officials. With his life in grave peril, he converted to Islam, much to the dismay of his followers.

The Muslim faith shares with Judaism the notion that world history is prophesied to someday draw to a close, and with Zoroastrianism, strong beliefs in heaven and hell. In the Koran, the Islamic holy book, there are lengthy descriptions of the pleasures waiting for the devout, and the torments reserved for nonbelievers and other sinners. It is emblematic of the similarities between the Muslim and ancient Persian concepts of last judgment, that Islam took hold in Iran so rapidly and effectively.

Perhaps the most powerful religious description of the end of the world (one also probably influenced by Zoroastrianism) is the Christian account of apocalypse in the book of Revelation. Revelation is one of the most over-interpreted and little understood parts of the Bible. Its horrific imagery of the catastrophic end of the world (some of it loosely borrowed from the book of Daniel) is so complex that it can be read in hundreds of ways—from a realistic chronicle of mundane political events to an allegory of the struggle between good and evil in heaven. The timeline that it follows is similarly vague; it does not specify whether its enumerated events are supposed to have taken place in the years immediately after the crucifixion of Jesus, one thousand years afterward, today, or during a period far in the distant future. For this reason, devotees throughout the ages have sought to interpret it in the manner most concordant with their own beliefs.

For the very first Christians, the prophetic passages of Revelation must have possessed a terrifying (but spiritually fulfilling) immediacy. As years went by, however, and the world did not experience the tribulations described in that book, the theological developers of nascent Christianity faced the task of deciphering its enigmatic timeline. If Revelation did not refer to an imminent future, they wondered to when exactly it did apply.

Critical to their endeavor was understanding the nature of the eschatologically significant *millennia* it describes. In one passage, it details a thousand year period in which Satan is bound in a bottomless pit. In another, it describes a thousand year interval in which the righteous reign with Christ.

Like detectives trying to solve a mystery by sifting through available evidence, many theological interpreters viewed these passages as critical clues in the understanding of God's ordained future chronology of the human race. Using these lines, along with other biblical descriptions, they attempted to pin down precisely when the world would end. Their prognostications took on heightened poignancy during times of great hardship, particularly during the bleakest days of the Middle Ages when disease swept relentlessly through Europe. The words of Revelation tolled like somber church bells in the hearts of those faced with untold devastation and unbearable sorrow.

CHAPTER TWO

THE PLAGUE OF FLORENCE

MEDIEVAL PERSPECTIVES

*In the year of Our Lord 1348 the deadly plague broke out in
the great city of Florence, most beautiful of Italian cities.
Whether through the operation of the heavenly bodies or
because of our own inequities which the just wrath of God
sought to correct, the plague had arisen in the East some
years before, causing the death of countless human beings.
Neither knowledge nor human foresight availed against it . . .
in spite of everything, toward the spring of the year the
plague began to show its ravages in a way short of
miraculous. . . .
Between March and the following July it is estimated more
than a hundred thousand human beings lost their lives
within the walls of Florence, what with the ravages
attendant on the plague and the barbarity of the survivors
toward the sick. Who would have thought before the plague
that the city had held so many inhabitants?*

—GIOVANNI BOCCACCIO (*Decameron*)

St. Augustine of Hippo

The Middle Ages in Europe are known as a time of great apocalyptic movements, arising among masses panicked by unbridled disease and famine. As waves of pestilence swept through lands unfamiliar with sanitary precautions, millions of Europeans met gruesome, untimely deaths. Those who desperately clung to life were paralyzed with anxiety. They were facing forces over which they had no control. No one knew when the next plague would break out and what its repercussions would be. This fear of a cataclysmic future served to fertilize myriads of doomsday cults. Painting their gruesome imagery using a palette of unspeakable horror—the vivid descriptions delineated in the book of Revelation—they portrayed, to the weary masses, scenes of impending universal calamity as deadly as the plague or even worse.

One cannot, however, characterize the whole of the Middle Ages as an age of apocalyptic terror. Fear of armageddon rose and fell, again and again, as each tide of disease crested and subsided throughout the lands of Europe. Apocalyptic fervor also waxed and waned due to the coming and going of other hardships—natural catastrophes such as earthquakes, volcanic eruptions, fires, floods, and droughts, as well as social problems such as warfare, crime and poverty.

The early Middle Ages, the fifth through ninth centuries A.D., are thought to have been generally far more peaceful and orderly than later periods. During the first part of the medieval epoch, the population of Europe was relatively low, mainly spread out through the countryside rather than concentrated in cities and towns. Therefore, disease did not spread very rapidly, certainly not as fast as it would several centuries later when Europe became more crowded. Anxiety was much lower and social stability higher than it would be later. For these reasons, faith in a growing and organized Catholic Church, rather than fear of future calamity, ruled the day.

With this increasing social order came the church hierarchy's desire to cast the book of Revelation—with its descriptions of coming calamity, the most radical book of the Bible—in a more conservative light. The first theological attempt to do so, even before the Middle Ages began, was performed by Origen, a third century scholar of classical Greek philosophy, born in Alexandria. Origen suggested the prophetic segments of the New Testament should be applied to the individual, rather than to the world as a whole. Instead of a literal chronicle of the future, he believed that these enigmatic writings should be viewed as an allegory, pointing the way to personal redemption. Thus, true believers, who have found the way to righteousness, need not fear global apocalypse. Stimulated by Origen's ideas, the church began to consider seriously metaphorical interpretations of New Testament prophecies.

Without a doubt, the religious scholar who shaped early medieval belief to the greatest extent was St. Augustine, Bishop of Hippo. He particularly helped to adapt Catholic philosophy to the growing conservatism of the time. Bringing order to seeming chaos, Augustine molded the disturbing imagery presented in Revelation into solid, soothing theological doctrine. At a time when numerous hands were attempting to mold Christianity, Augustine's extraordinary gift of rhetoric established him as the artisan of an astonishingly resilient philosophical framework upon which clerical thought was hung during the Middle Ages.

Augustine was born in 354, in Tagaste, Numidia (now Souk-Ahras, Algeria), a Roman province of North Africa, to a pagan father and a Christian mother. In spite of his mother's considerable influence (for which she was later canonized as Saint Monica), Augustine embraced Christianity only after years of experimentation with other faiths. As a young man, he was particularly drawn to the Manichaeist religion, a popular system of beliefs at the time. Manichaeism, the doctrine of the fourth century Persian prophet Mani, represented a unique combination of Zoroastrianism, Buddhism and Christianity. After immersing himself in this faith, Augustine eventually found that Mani's teachings did not address all of his spiritual and emotional needs. Dabbling in several other mystical religions, he finally became a convert to Christianity. Impressing church officials with his gift of rhetoric, Augustine was consecrated Bishop of Hippo, North Africa in 395.

Augustine was a remarkably prolific author. He wrote 118 treatises, including *The Confessions*, a candid spiritual autobiography, and *City of God*, a comprehensive interpretation of the Bible that became a pillar of Catholicism. Much like the case of a prosecutor becoming a defense attorney, his years outside of Christianity gave him a keen sense of ways of anticipating and then countering critiques of the faith.

In his position of critic-turned-supporter, Augustine was very much aware of the puzzling aspects of the book of Revelation. In *City of God*, he urged Christians not to interpret the last book of the New Testament—especially its references to the thousand year reign of Christ—in too frivolous a manner. "Some Christians," he decried, "... construe the passage into ridiculous fancies."[1]

Augustine was alluding to a series of verses that have spurred on the prognostications of doomsday prophets for generations:

> And I saw an angel come down from heaven, having the key of the bottomless pit and a great chain in his hand. And he laid hold on the dragon, that old serpent, which is the Devil, and Satan, and bound him a thousand years, and cast him into the bottomless pit, and shut him up, and set a seal upon him, that he should deceive the nations no more, till the thousand years should be fulfilled: and after that he must be loosed a little season. . . .
>
> Blessed and holy is he that hath part in the first resurrection: on such the second death hath no power, but they shall be priests of God and of Christ, and shall reign with him a thousand years. And when the thousand years are expired, Satan shall be loosed out of his prison, and shall go out to deceive the nations which are in the four quarters of the earth, Gog and Magog, to gather them together to battle . . .[2]

One popular early Christian interpretation of this biblical passage asserted that the end of the world was imminent. Millenarians, believers in an impending time of apocalypse, felt that they could combine the predictions of this portion of Revelation with the chronology suggested by other biblical references to foretell precisely when doomsday would arrive. By synthesizing the account in Genesis that the world was created in six days, with the statement in Peter, "One day is with the Lord as a thousand years, and a thousand years as one day,"[3] they established a time scale for history. Estimating, by these calculations, that it would take a total of six thousand years for the human chronicle to unfold, and reckoning that Earth was already thousands of years old, they predicted that the end of time could not be far off. The year of universal destruction, based on these estimates, was initially calculated to be A.D. 500. Later, however, after that year came and went without event, revised opinion put it at A.D. 801. Over the centuries, predictors of apocalypse have forecast new dates of doom again and again to explain away unmet prophesies.

The millenarians believed that the final millennium of history (in which they thought they were living) consists of a time in which the goodness of God keeps the evil of Satan in check. Very soon, they preached, Satan will be released from his prison and wreak absolute chaos upon the world. The horrible events mentioned in Revelation—the rain of fire, crashing of stars, plagues of locusts, etc.—would transpire, one after another, until Earth as we know it was destroyed.

Augustine countered these forecasts, which he found to be unnecessarily alarming, with an alternative, looser, model of time. The period of one thousand years, he implored, should not be taken literally as the fixed duration of a particular biblical era. Rather, he suggested, it should be considered in a purely figurative manner. To him, as the cube of the number ten, "one thousand" seemed to embody perfection. Biblical references to that figure, he felt, were meant as allusions to the perfect nature of the totality of human history. Thus, for the Bishop of Hippo, "one thousand years" signified "the whole of man's days on Earth."

By peeling away the literal implications of the chronology suggested in parts of Revelation, Augustine exposed what he saw as the metaphorical kernel of the text, one that lent itself well to mainstream church teachings of the time. He viewed the last book of the Christian Bible as a testament to the purity of those who honor Christ and belong to his church. Satan was free, he felt, to muddle the hearts and minds of those outside of Christendom, but could not affect those within its fortress. Thus, in Augustine's opinion, Revelation served as a tribute to the protective power and longevity of the church, by drawing attention to the horrific alternatives.

Augustinian doctrine thereby served to dissuade Christians from ruminating on when and how the apocalypse would occur. Rather, it urged believers to focus their spiritual energies on embracing the community of the church, and to follow its practices as a bulwark against evil. The devout—those who reign with Christ and share his love and protection—need not fear the end of the world, and should waste little time worrying about the ghastly images of destruction in Revelation.

Augustine's words of caution served well the interests of the church of his time. Although Christianity began as a small alternative movement, by the fourth century it had become the official creed of the Roman Empire and a dominant force around the world. Its self-preserving interests were in allaying, not fanning, the fears of the populace. Church officials knew that only anarchy would ensue if apocalyptic anxiety were to get out of hand. Thus Augustinian belief very quickly became adopted as orthodox doctrine.[4]

Augustine's calming words—coupled with the stability of the age—served to prevent millenarian thought from having much impact during the early Middle Ages. The common folk, locked into lives of hard work in the field, and simple devotion in church, saw no need to panic in the face of prophesied doom. Only a few rogue individuals, who happened to have fallen through small tears in an otherwise tightly woven social network, had reason to preach of disaster.

With prosperity in Europe, however, came a swelling of population that ironically led to a gradual undoing of the calm. Growing numbers of people meant larger cities, substantial overcrowding, rising homelessness, devastating poverty, and virulent disease. As these social problems took their toll, talk of coming doomsday built from a background whisper to a rising crescendo.

The Year 1000

In our privileged historical position to experience the turning of the third millennium, we might naturally wonder what happened during the last such opportunity: the advent of the year 1000. Considering biblical references to thousand year spans, particularly in

Revelation, one might imagine that the medievals viewed the coming of that year—traditionally the thousandth since the coming of Jesus—very seriously. But that need not be the case. One might conversely conjecture that Augustinian cautions about the dangers of millenarianism would serve to temper all sense of panic.

What actually occurred in the year 1000 is a matter of debate. Numerous contemporary scholars have studied that year, examining whether it was a time of great trepidation or a period just like any other. Original sources chronicling the history of that year are so rare, thus there is considerable controversy regarding this question. Most current medieval historians feel that 1000 came and went with little impact. They point to the sparsity of direct evidence of apocalyptic fervor during that period, and dismiss secondhand accounts as being prone to exaggeration.

However a few prominent scholars, such as Richard Landes, Director of the Center for Millennial Studies at Boston University, argue the opposite. They believe the writings available from that era, reporting harbingers associated with the coming of the new millennium, are generally valid. Landes, who has launched an intensive study of the year 1000, notes the overwhelming confusion throughout Western Europe at that time, calling it "the mother of all apocalyptic moments."[5]

One of the primary original sources from this period is a startlingly vivid account by Radulphus Glaber, a French monk. Glaber, who was put into a monastic school by his uncle, led the life of a vagabond, wandering from one monastery to another. During the course of his travels, he kept a careful record of his impressions. His written journal, and presumably his life, seems to have ended at Cluny around 1044.

Glaber's writings frame the years around the turn of the second millennium as a time of absolute chaos. His portrait of that age pictures Christians bracing themselves for apocalypse while they witnessed a terrifying chain of ominous tragedies. One after another, a series of natural disasters brought their houses, families, marketplaces, and churches all to states of ruin. The petrified masses became increasingly certain with each catastrophe that their days on Earth were numbered. This string of cataclysmic events began, as

Glaber relates, with the sudden eruption of Italian volcano Mt. Vesuvius:

> In the seventh year before [the year 1000] Mount Vesuvius, which is also called Vulcan's cauldron, gaped far more often than his wont and belched forth a multitude of vast stones mingled with sulfurous flames which fell even to a distance of three miles around. . . . It befell meanwhile that almost all the cities of Italy and Gaul were ravaged by flames of fire, and that the greater part even of the city of Rome was devoured by a conflagration. . . . At the same time a horrible plague raged among men, namely a hidden fire which, upon whatsoever limb it fastened, consumed it and severed it from the body. . . . Moreover, about the same time, a most mighty famine raged for five years throughout the Roman world, so that no region could be heard of which was not hunger stricken for lack of bread, and many of the people were starved to death. In those days also, in many regions, the horrible famine compelled men to make their food not only of unclean beasts and creeping things, but even of men's, women's, and children's flesh, without regard even of kindred; for so fierce waxed this hunger that grown-up sons devoured their mothers, and mothers, forgetting their maternal love, ate their babes.[6]

Parts of Glaber's account, especially the section about cannibalism, sound rather far fetched. It is hard to imagine the taboo against the consumption of human flesh being violated on such a major scale. Perhaps Glaber was engaging in a bit of hyperbole at that point of his narrative in order to emphasize the misery of his times. Since no other texts corroborate nor contradict the period Glaber is describing, it is hard to sort fact from fiction.

Indeed, Cambridge historian G. G. Coulton, one of the principal modern scholars of that period, cautions in a translation of Glaber's text not to draw "exaggerated deductions" from that work.[7] Coulton questions the extent to which the year 1000 was special, and emphasizes that "it was not only at and about this date that our forefathers expected strange events; the medieval mind was perpetually haunted by the expectation . . . that the end of all things was at hand."[8]

In spite of Coulton's admonition, a few contemporary scholars have taken the liberty to embellish upon chronicles that have been passed down from that period. Consider, for example, this second hand account of the dreaded final hours before 1000, as told by writer Richard Erdoes:

> On the last day of the year 999, according to an ancient chronicle, the old basilica of St. Peter's at Rome was thronged with a mass of weeping and trembling worshipers awaiting the end of the world. This was the dreaded eve of the millennium, the Day of Wrath when the earth would dissolve into ashes. . . . As the minutes passed and the fateful hour was about to strike, a deathly silence filled the venerable basilica. . . .
>
> But when the fatal hour passed and the earth did not open to swallow up church and worshipers, and when no fire fell from heaven burning everything alive to ashes, all stirred as if awaking from a bad dream. Then everybody breathed a sigh of relief amid much weeping and laughing. Husband and wife, servant and master embraced each other as friends and exchanged the kiss of peace. Then all the bells of St. Peter's, of the Lateran, of the Aventine, of every church upon the Seven Hills of Rome began to ring, praising the Lord as with one single voice. The bitter cup had passed, the world was like reborn and all humankind rejoiced, as related by many ancient chroniclers.[9]

This sensational account, though based on remarks by Glaber and other medievals, seems to be an example of the sort of embellishment that Coulton warns about. Yet because we cannot return to those times and find out what actually occurred one thousand years ago, we might only hope that new primary sources appear and help clear up the matter.

Joachim of Fiore

Though the question of how much doomsday panic gripped the masses in the year 1000 is a matter of controversy, present-day historians are much more certain of the sizable extent to which it affected thirteenth century European believers. In the intervening centuries, poverty, overpopulation and disease had risen considerably. These social ills led to a great increase in the number and influence of apocalyptic movements.

Modern scholars readily explain how the conditions for this mass hysteria developed. The populace in Europe had grown far beyond what the age-old agricultural system could possibly support. With nowhere else to turn, thousands of peasants crowded into newly-established urban centers. Some of them found work, but a great number were unemployed. In short order, a permanent underclass developed, with little constructive role to play in society. This group of beggars, vagabonds, and other outsiders had nothing to lose—and a sense of solidarity to gain—by joining fringe movements. Whenever a charismatic leader required a cadre of loyal followers, he need only look to these urban misfits for support.

The lure of apocalyptic cults grew even stronger during times of natural disaster. Whenever the lives of the displaced peasantry were disturbed by any kind of unforeseen catastrophe, their collective rage often became channeled into groups preaching that the world would soon end. As British historian Norman Cohn relates:

> Any disturbing, frightening or exciting event—any kind of revolt or revolution, a summons to a crusade, an interregnum, a plague or famine, anything in fact which disrupted the normal routine of social life—acted on these people with peculiar sharpness and called forth reactions of peculiar violence. And one way in which they attempted to deal with their common plight was to form a salvationist group under a messianic leader.[10]

St. Augustine's writings had steered devout Christians away from looking at Revelation as a predictive text. Because of his influence, for many centuries belief in a catastrophic conclusion to history won little support among church officials. When in the thirteenth century, apocalyptic movements were on the rise again, these groups turned to another Catholic philosopher for guidance, Joachim, the Abbot of Fiore.

Joachim of Fiore was born in Calabria, Italy, in 1145. After many years of studying the Bible, he began to suspect hidden meaning buried within its text. Some time between 1190 and 1195, he developed a new biblical approach that, in his opinion, enabled him to predict the future. Joachim's novel interpretation found numerous parallels between the Old and New Testaments—relating the stories of Adam and Abraham to the chronicles of Jesus and the apostles. Extrapolating forward in time—toward the time of his writing—he postulated further parallels between the age of Jesus and the Christian monastic period.

The essence of Joachim's work is that the history of the world can be divided into three distinct epochs, each corresponding to a member of the Holy Trinity. The Holy Trinity (Father, Son, and Holy Spirit) is considered by Catholics to embody the three principal aspects of God. In Joachim's system, the first part of the biblical chronicle—lasting from Adam until the generation before Jesus—can be represented by the Father, and is known as the "Age of Law." According to the Abbot of Fiore, this initial stage of history was a time in which the heavy hand of divine justice ruled the world.

Following the "Age of Law," in this scheme, was a time of heavenly mercy, ruled by the Son, and called the "Age of the Gospel." The "Age of the Gospel" consisted of the historic period that started with the birth of Christ, and culminated in the establishment of the church order by Roman Emperor Constantine.

Finally, Joachim delineated a third and ultimate period of history: the "Age of the Holy Spirit." In contrast to earlier times, this era will be characterized by universal joy and love. Lasting until the end of time itself, it will be a period in which everyone lives in a state of spiritual bliss. The model for life during this era will be that of a monastic order, where all are servants of God. Humankind will spend their days chanting and praying—reveling forever in the beauty of divine creation.

Based upon his outline of history, Joachim estimated when he thought the mundane world would be destroyed, and the new spiritual age would begin. Using the genealogy detailed in the Gospel of Matthew as a guide, he reckoned that there had been forty-two generations from Adam to Jesus. Since he believed the Old and New Testaments parallel each other, he concluded that there would be forty-two additional generations between the time of Christ and the end of the world. Human history would draw to a close sometime shortly after 1200, he calculated; most likely it would end around 1260.

Joachim predicted that the last few years before the culmination of history would be a time of utter chaos. An evil king would gain phenomenal power throughout the world, playing the role that the New Testament refers to as the "antichrist." Through his sinister doings, he would wreak havoc upon humankind and bring about the absolute destruction of society. Only after the human race became immersed in the waters of total devastation would it emerge pure and holy, and ready for the Age of the Spirit.

The Abbot of Fiore was hardly a radical. His work was sanctioned by three popes, who found great worth in his biblical interpretation and saw him as no threat. Although his conclusions about the shape of things to come were drastically different from those of Augustine, Joachim did his best to offer tribute to his more conservative philosophical predecessor. He certainly had no desire to overturn existing religious institutions and establish a new way of thinking.

Yet, in spite of Joachim's modest nature and limited goals, his writings met the collected expectations of the masses like the summer sun on dried leaves. As his predicted year of disaster—1260—approached, a bonfire of dread began to spread across the European continent. Engulfing thousands of self-perceived sinners who feared God's wrath, this conflagration of apocalyptic terror left few places untouched.

Picture a parade of tormented men, their backs bloodied by self-inflicted torture, praying fervently that their souls would be saved. Carrying banners by day and lit candles by night, they would march through the streets of towns and villages, until they arrived at a main square, in front of a church. There they would stop and enact their horrifying public ritual of punishing themselves for their sins. Whipping themselves brutally for hours while chanting slogans of redemption, they hoped their earthly penance would excuse them from heavenly punishment. Invariably, townspeople of all classes and occupations would join the group in a frenzy of repentance. Such was the vicious scene in much of Europe during the dark days of the thirteenth century.

Though there are records of individual cases of religious self-flagellation dating back to the eleventh century, the organized flagellant movement was launched in 1260 by a hermit of Perugia, Italy. The strange crusade spread like wildfire through the Italian heartland and then ultimately north to Germany and other parts of Europe. Terrorized by famine in 1258 and plague in 1259, as well as constant interregional warfare, it seemed that those crowded into the towns and cities of central Italy were ripe for a movement that would exploit their apprehension about the future. Self-flagellation was embraced as a way of warding off doom by identifying with the sacrifice of Christ. Those who marched in flagellant processions hoped to place themselves among the elect that would survive the coming apocalypse and enjoy the peaceful years of the coming Age of the Holy Spirit.

The flagellant movement continued in Europe for several centuries, its numbers rising during turbulent times and falling during periods of relative calm and prosperity. In some countries, such as Italy, practitioners were welcomed as members of the organized church; in others, such as Germany, they were persecuted as heretics. Their persecution, which they saw as martyrdom, made them even holier in their own eyes.

Generally, the attitude of church officials toward the flagellants was supportive as long as they did not threaten to usurp Christian ritual authority. Once the lay members of flagellant groups began performing sacraments, hearing confessions, and claiming powers of absolution, they lost their official support. When they began to attack the Jews, whose welfare as a "biblical people" was considered part of Christian charity, they antagonized the church even more.

In 1349, Pope Clement, who initially sympathized with the flagellants but later realized the severity of their threat, issued a papal bull decreeing they were to be banished from all good Christian communities. In his edict, he cited their persecution of Jewish, and even some Christian, groups as among their heresies.

Clement acted just in time. By then the movement had grown into an overwhelmingly powerful force. According to medieval reports, flagellant parades numbering in the tens of thousands were quite common during that era. Reportedly, a procession in the German town of Constance drew about forty-two thousand devotees, and one in Brabant attracted over eighty thousand. If it weren't for Clement's decisive action, the destructive capacity of the flagellants would have increased beyond hope of containment.

It took an especially terrible event to cause such a terrifying crusade to grow so rapidly. The self-flagellation movement reached its peak during the time of the Black Death in the fourteenth century, a period of unbearable hardship and untold misery due to natural causes that fortunately has never been equaled.

The Black Death

The worst catastrophe in recorded human history was produced by the tiniest of agents: an especially virulent type of bacteria that, in a several year period, unleashed on more than one-third of Europe and Asia the experience of gruesome, untimely deaths. No one could have foreseen the ghastly toll this pernicious illness would have upon the world.

The plague bacillus, *Yersinia pestis*, first identified by scientists in the 1890s, is a short bacteria that is said, under the microscope, to resemble a safety pin. It is carried primarily by fleas buried in the fur of wild rodents (rats, mice, and squirrels), and is transmitted to humans by flea bites. It is then spread from victim to victim by means of the droplets expelled through coughing.

The most common form of the illness is called the bubonic plague. This type of plague, named for the *buboes,* or swellings, it creates, has an incubation (asymptomatic) period that can range from a few hours to twelve days. If left untreated, it brings on chills and a high fever, accompanied by enlarged lymph nodes in the groin and armpits. These buboes, about the size of eggs, are hard, tender, and unsightly. They serve as visual harbingers of even more dreadful symptoms to come. Within a few days, hapless sufferers typically lose their coordination, their reason, and their demeanor. They lapse into states of utter delirium, and then finally, in most cases, they die.

A less typical, and even deadlier, form of the disease is known as the pneumonic plague. Its main manifestation is a wretched cough that develops within twenty-four hours of being exposed and produces a sputum that grows bloodier and bloodier. As the lungs fill with fluid, breathing becomes more and more difficult, until painful death ensues. Other symptoms of this ghastly affliction include foaming at the mouth, fever, chills, rapid heartbeat, and headaches. Often pneumonic plague appears in the midst of a bubonic plague epidemic; the two illnesses are closely related.

Today, plague can be cured with tetracycline and other common antibiotics. If caught early enough it is rarely fatal. Only in Southeast Asia, and a few other underdeveloped parts of the world, does it represent a serious health problem. But in the Middle Ages it was a terrible scourge that granted no reprieve, killing many more people than all of the wars of the twentieth century combined.

It can be argued that the reaction to the horrors of the plague was the greatest influence upon the shape of late medieval life and thought. Never before were so many convinced that the world was about to collapse around them. The style of thinking of the times became focused on mere survival. Faith and prayer provided the only antidote to despair. Divine guidance was seen as the sole means of bridging the crumbling ruins of an impossible present and stepping forward into a saner future.

Europe in the 1340s was a paper fortress lying in the path of the foul winds of fate. Its crowded cities with abysmal sanitation provided ideal breeding grounds for disease. Living quarters were cramped, providing ample opportunity for microbes to pass from person to person. Poorly stored food became breeding grounds for insects and vermin. Drinking water was often taken from the same source in which raw sewage was dumped. Worst of all, people almost never bathed or even washed their hands. It is no wonder that a deadly microorganism, such as the bubonic/pneumonic plague, could spread so effectively and decimate European society so quickly.

The origins of what came to be called the Black Death (because of characteristic black patches formed on the skin of its victims) are somewhat uncertain. Scientists today believe that the outbreak started around 1346 somewhere in Central Asia; speculations of its birthplace range from Mongolia to India, to the Sinkiang region of western China. At that time, millions of rats, infested with infected fleas, were driven from their natural habitats by an acute food shortage and forced to forage among human dwellings. No one knows what caused the sudden scarcity of supplies. Ancient Chinese accounts speak of a series of violent earthquakes that rocked Central Asia in the mid-1340s. Perhaps these natural convulsions spurred on the famine that drove the rats from their homes. Some modern researchers have argued instead that favorable weather during that period caused an overpopulation of vermin that generated the food shortage. At any rate, medieval chronicles report of devastating pestilence in Asia soon after villages were besieged by armies of rodents.

Rapidly, the plague spread across the lush mountains and valleys of China, throughout the monsoon-soaked provinces of India and over the sun-baked steppes and plateaus of Asia Minor. Records show that the Chinese population fell from 123 million in 1200 to 65 million in 1393, probably due to plague and subsequent famine.[11]

Rumors of a terrible Asian infestation reached European soil well before the actual disease arrived there and began taking its ghastly toll. Once the plague came to Europe, it proved even deadlier than anticipated. It preyed on that continent like a vampire, swiftly draining its lifeblood, and rendering it a pale carcass.

The first outbreak was in Italy. While on trade expeditions to eastern lands, Italian merchants contracted the horrible illness, and subsequently brought it back to their homeland. In 1347, a group of Genoan sailors returning from the Crimea found themselves racked with high fever and coughing up blood. Their groins and armpits swelled up painfully and grotesquely—physical symptoms soon recognized as unmistakable harbingers of impending demise.

Accompanying them aboard the death ship as it sailed homeward were hundreds of plague-carrying rats. When the craft arrived in the port of Messina, Sicily, the human victims and rodent carriers soon mingled with and infected the local population. In short order, Messina became a town under siege, attacked by a biological monstrosity of gargantuan proportions.

From the crowded port towns, the plague rapidly spread through the Italian countryside, and then north into Germany, France, and the bulk of an ill-fated Europe. It even crossed the channel into England and the Mediterranean into North Africa. From the largest cities to the smallest hamlets there was no respite from the invisible stalker.

Faced with impending mortality, friends and family members turned on one another like sworn enemies. Parents would abandon their own ailing children, throwing them out of their own homes because of fear of contamination. Priests would refuse to comfort the dying or to offer last rites, and undertakers would refrain from burying the deceased. Blackened bodies would pile up in fetid mountains, lending evil aspect to the ruined cities. Those lucky individuals who had the chance—the nobility, for example—would cloister themselves as far away as they could from plagued communities. Once they found refuge, they barred anyone from entering their shelters.

The most famous description of the unsettling effects of the plague upon the medieval psyche is the *Decameron*, penned by renowned fourteenth century Florentine poet Giovanni Boccaccio. With a distinct mix of pathos and humor, Boccaccio pictures a group of upper class youth fleeing to a country villa in the hope of avoiding infection. While nestled in their retreat, they occupy their time gossiping and exchanging humorous tales, waiting for the age of pestilence to end. Meanwhile, their less fortunate, lower-class fellow citizenry, stuck in Florence and other besieged Italian cities, succumb in great number to the ghastly malady. Boccaccio describes the anguish of those left behind to suffer:

> It was not merely a question of one citizen avoiding another, and of people almost invariably neglecting their neighbors and rarely or never visiting their relatives, addressing them only from a distance; this scourge had implanted so great a terror in the hearts of men and women that brothers abandoned brothers, uncles their nephews,

sisters their brothers, and in many cases wives deserted their husbands. But even worse, and almost incredible, was the fact that fathers and mothers refused to nurse their own children, as though they did not belong to them.

Hence the countless numbers of people who fell ill, both male and female, were entirely dependent upon either the charity of friends (who were few and far between) or the greed of servants, who remained in short supply despite the attraction of high wages out of proportion to the service they performed . . . A great many people died who would perhaps have survived had they received some assistance.[12]

Of course, at a time well before modern medicine took root, no one had an inkling as to the cause of the great catastrophe. Many considered it punishment for human sin—an act of divine retribution—representing, perhaps, one of the cataclysms foreshadowed in Revelation. Indeed the last book of the New Testament warns that during the end times an angel will release a plague of locusts, attacking those hapless souls who do not bear God's seal upon their foreheads.

Believing Judgment Day was at hand, many took to the streets, filling the ranks of fear-crazed fanatical movements. The flagellants in particular saw a hefty increase in the size and zealotry of their membership. Jews and other outsiders were selected as victims of bloody pogroms, scapegoated for their alleged use of "black magic" to spread illness. Apocalyptic cults saw themselves as purging the human race of sin and sinners, believing that only religious purity would enable God to call off his horsemen of doom. Doing nothing to hasten the end of pestilence, their efforts merely amplified the misery of those who somehow survived the death of their loved ones.

Although there were to be many other outbreaks of plague around the world, the great epidemic ended around 1350 for most of Europe. Historians have speculated that by then more than twenty-five million Europeans out of a total population of about eighty million, and perhaps more than seventy-five million people worldwide, out of a population of five hundred million, succumbed to the horrible malady in its many forms. To try to fathom the enormity of the loss that occurred, picture the impact on the United States of the sudden decimation of the states of California, Texas, Illinois, and New York.

One indication of the toll taken by the Black Death on European citizenry (little is known of its affect on Asian citizenry) was an enormous post-plague labor shortage. This manifested itself in a sharp increase in the price of workmanship. Workers lucky to have survived the plague demanded high wages for their services, compared to what they had charged earlier.

Aside from the shortage of workers, it is surprising how quickly civil society returned to normal in Europe, after the passing of such an unimaginable cataclysm. Within decades, Europe entered an age of general prosperity, buoyed by rising salaries and a bolstered standard of living, as well as the temporary relief of urban overcrowding (albeit through a gruesome agent).

Jewish citizens and other victims of massacres by flagellants could share in the general sense of relief. Pope Clement's 1349 edict, and the return of civil order following the abatement of the plague, helped to squelch the flames of the self-flagellation movement. Faced with the concerted opposition of the church, as well as of civil society, the flagellant crusade largely died out in the 1350s. Hangings and beheadings of flagellants, authorized by various church tribunals, dealt a critical blow to the movement, until it terrorized Europe no more.

Danse Macabre

With the horrors of the Black Death eased and the flagellant movement quashed, apocalyptic hysteria, at least in its overt forms, became less and less of a force in Europe. The years of panic became a fading memory, as European society regrouped and moved on. Yet somewhere in the collective subconscious, visions of those unsavory times were etched.

One of the lingering manifestations of the anxiety caused by the plague years was the peculiar medieval art form known as the *danse macabre* (the dance of death). This expression refers to the morbid depictions of the power of death that are omnipresent in the paintings, writings, and other cultural artifacts that date back to that period. In a typical rendition, Death, depicted as either a corpse or a skeleton, is shown coming to steal away young and old, rich and poor, sinners and saints, the ugly and the beautiful alike. In the face of mortality, this vision suggests, no one should be vain enough to esteem his or her own body, for it will soon enough be eaten away by worms and maggots. Only the eternal soul should be valued.

"Danse macabre" is an expression that originated in France. "Macabre" probably derives from a corruption of "Maccabee," referring to the biblical tale in the second book of Maccabees of seven martyred brothers. The first known use of that term was in a 1376 poem by Jean Le Fevre. By the end of the fourteenth century, the gruesome concept formed the theme of a play performed in Normandy, and the subject of a morbid dance song.

Imagine going out for a night on the town, dressing in black, putting on white, skeletal make-up, and chanting somber tunes about death and dismemberment. The fashions of 1980s New Wave dance clubs would have paled in comparison to those of the late fourteenth century. London and New York punks could never match the ghastly spirit of those medieval French artists and performers.

Probably the most famous relics of the danse macabre period are the grim, but ornate, series of woodcarvings found on European church walls and other public structures and can be seen in books from the late Middle Ages. These relics were carved in the fifteenth and sixteenth centuries, well after the dance of death had first taken hold as a cultural symbol. Particularly well known examples of this genre were fashioned by Albrecht Durer (1471–1528) and Hans Holbein the Younger (1497–1543).

The Final Plague

Considering the untold horrors of the plague years, it is incomprehensible to ponder what it would be like if such a horror ever were to wreak the Earth again. Modern medicine can now cure the plague bacillus (*Yersinia pestis*) itself, but might there be comparable or even worse afflictions for which there is no counteragent? Might a new, extraordinary lethal analogue of the Black Death even wipe out the human race?

In the past half century, humankind has acquired a frightening dependence on the use of antibiotics to ward off illness. Like terrified medieval nobility barricading themselves in imposing castles to avoid contact with lethal ailments, we shelter ourselves in what we hope are impenetrable pharmaceutical fortresses, hoping that no microorganism will prove strong enough to break through. Yet in spite of this reliance on modern medicine, bacteria and other microbes are challenging—and often bypassing—pharmaceutical barriers with disturbing frequency. Immunologists, such as Jeffery Fisher of the Helicon Foundation, and scientific experts, such as journalist Laurie Garrett, a fellow at the Harvard School of Public Health, are warning that it is only a matter of time before a new epidemic sweeps the earth on the scale of the Black Death (or maybe even worse) for which antibiotics and other known remedies will prove ineffective. They point to AIDS as a harbinger of a coming age of menacingly adaptable, highly resistant diseases that are tough or even impossible to eradicate.

Sir Alexander Fleming of Scotland, who in 1928 discovered penicillin, the first modern antibiotic, warned in 1942 (when it was first being marketed) that bacteria might eventually develop a means of resisting the drug. In spite of Fleming's admonition, the use of antibiotics became ubiquitous as a precautionary measure, even when its application was hardly warranted. "Miracle drugs" became introduced into the food chain by farmers hoping to boost farm profits by medicating healthy farm animals to ward off possible infection.

Meanwhile, as the great Scottish biologist predicted, whenever scientists introduced a new type of antibiotic, strains of bacteria emerged against which that drug was ineffective. The mechanism by which this natural resistance developed could be well explained. Through the random action of mutation, new varieties of bacteria appeared—with some resisting a given antibiotic better than others. Over time, the Darwinian process of natural selection filtered out those bacteria with low resistance and favored those with high resistance against the drug in question. Ultimately, entire strains evolved which evaded capture by that agent.

As certain bacteria "learned" to counteract particular types of antibiotics (more precisely, those that didn't do so, died out), they passed along this information to others, through genetic structures called plasmids, until entire categories of bacteria became resistant. During a twenty-year period, medical studies showed that the effectiveness of sulfa drugs (pharmaceutical agents with antibacterial properties) against some ailments *declined by over 50 percent!*

Fisher points out three ways bacteria have developed counterattacks against antibiotics.[13] The first line of defense, called drug inactivation, involves producing an enzyme designed to disable the antibiotic. Biochemists have retaliated by developing new types of antibiotics specifically formulated to resist this bacterial enzyme. Fisher predicts bacteria will soon find a way of getting around that hurdle, and that new resistive measures will prove necessary.

The second means of bacterial resistance is an altered target site. This entails microorganisms to change their own structures by means of mutation until antibiotic agents can no longer bind to them. Like a blinded predator groping through a forest futilely attempting to capture its prey, the drug can no longer "see" the altered bacteria and passes over without squelching them. The only way for pharmacists to block this avenue of defense is to design an improved antibiotic—a daunting task.

Finally, the third method by which bacterial villains can avoid the pharmacological posse involves bootlegging new biochemical substances. By manufacturing novel enzymes—ones that antibiotics are not designed to block—the bacteria can continue to carry out their dastardly work. Metabolic bypass is one example of this method. This occurs when the effectiveness of sulfa drugs, which work by stopping bacteria from producing folic acid, is counteracted by a new bacterial enzyme that creates it in a different way.

To combat these bacterial defenses, Fisher offers a number of valuable suggestions aimed at increasing the long-term effectiveness of antibiotics.[14] First, and foremost, he advocates educational programs—in medical school and elsewhere—designed to inform physicians of the hazards of the overuse and misuse of antibacterial medications. These lessons should include a primer on how bacteria develop resistance and a compendium of what new resistant strains have arisen.

Fisher also endorses Harvard professor Dr. Jerry Avorn's notion of "lifestyle prescriptions." "Lifestyle prescriptions" are physicians' suggestions of nonpharmaceutical measures, such as taking vitamins or resting, that may help patients recover more quickly. Many patients demand drug prescriptions from doctors each time they have simple ailments, even in cases where no medicine can cure what ails them. For example, they come in with a cold and want a quick cure, even though there is none. Antibiotics are ineffective against colds and other viral infections. In those cases, Fisher states, some physicians prescribe antibiotics anyway, thinking they can't hurt, and that occasionally viral infections predispose to bacterial infections. However, the more antibiotics prescribed, the greater the chance that bacteria will grow familiar with them and develop resistance. Therefore, doctors should learn to analyze critically patients' desires for quick remedies, determine if these are legitimate, and if not, write out "lifestyle prescriptions" instead. By receiving such a script, patients might feel that their doctors are doing something practical—reducing their desires to beg for unnecessary medicines that may lead to the spread of resistant strains elsewhere.

A number of concerned groups, such as the Food Animal Concerns Trust (FACT), have pushed for the elimination of antibiotics in feed for healthy animals. Fisher agrees with this cause, and views it as a further means of reducing humankind's overuse of these drugs. As an additional way of combating bacteria, he urges the development of new vaccines and drugs to fight their resistance and overcome infection. Also, many existing vaccines and alternative treatments, such as the inoculation against pneumococcal pneumonia, an illness that often affects the elderly, are not used with enough frequency. If all of these measures are taken, he believes that the medical use of antibiotics will return to more reasonable levels, and that the battle against harmful bacteria will consequently be bolstered.

Though fighting disease-causing bacteria constitutes a critical front in the modern war, it is hardly the only one. AIDS and hepatitis in their various forms, two of the most devastating epidemics of the twentieth century, are caused by viruses, not bacteria. Ebola, another catastrophic illness that has plagued parts of Africa, is a viral infection as well. And the so-called "Mad Cow Disease," a brain-destroying affliction borne by cattle, is caused by neither bacteria nor viruses, but rather by a newly identified biological agent called a prion.

At present, AIDS has no known cure. Its causal agent, HIV, can often be kept under control for years by the use of "cocktails": prescribed drugs that complement each other in effectiveness. These medications, however, are very expensive and hardly constitute a permanent solution to the problem. Only a tiny minority of those dying of AIDS in third world countries (where it is prevalent) have access to such costly treatments. Vaccines designed to prevent HIV infection are currently being tested.

In 1976, tragic scenes from the *Decameron* replayed themselves all too vividly in central Africa. A hospital in Yambuku, northern Zaire (now Congo) was faced with an unknown ailment of devastating proportions. Patients, arriving in droves, were vomiting and defecating massive amounts of blood for no apparent cause. Their noses and gums were also bleeding. Moreover, they were going mad. The Ebola epidemic had begun, one that was to decimate numerous African villages and kill hundreds.

Laurie Garrett, writing of the illness in her instrumental book, *The Coming Plague,* describes the horrific situation faced by the hospital staff:

> Soon the hospital was full of people suffering with the new symptoms. Panic spread as village elders spoke of an illness, unlike anything ever seen before, that made people bleed to death. In Yambuku the Sisters (religious hospital staff) were close to the breaking point, not knowing the why, what, or how of the new disease.
>
> The horror was magnified by the behavior of the many patients whose minds seemed to snap. Some tore off their clothes and ran out of the hospital, screaming incoherently. Others cried out to unseen visitors, or stared out of ghost eyes without recognizing wives, husbands, or children at their sides.[15]

Virologists isolated the Ebola virus shortly after the initial outbreak. Discovering that it carries a high mortality they learned the virus was transmitted through bodily fluids. Severe bleeding was associated with the Ebola virus infection, and this manifestation contributed to a high mortality rate. They began to work on developing treatment. Eventually they hoped to generate a vaccine. Unfortunately, because of the extremely contagious nature of the virus, progress has been slow. Neither a vaccine, nor an effective treatment, seems forthcoming.

In Britain, starting about a decade ago, farmers began to realize that a strange new malady was killing off many of their cattle. Veterinarians dissected the brains of some of the felled cows and discovered they resembled sponges—full of clear holes called vacuoles. Because of this resemblance, the disease became known as bovine spongiform encephalopathy (BSE).

About the same time as the epidemic began, American neurologist Stanley Prusiner made a startling discovery, one that had strong bearing on the search for the illness' cause. He found a new disease carrying agent, called a prion (proteinaceous infectious particle), that is the transformed version of a cellular protein present in all forms of animal life, including humans. Prions are much smaller than viruses, bacteria, and fungi, the other major carriers of infectious disease. He discovered these particles could link up in chains, slowly wreaking havoc on the central nervous system. Moreover, prions can be transmitted from animal to animal through either inheritance or ingestion. He proved that scrapie (a sheep illness) and BSE are each caused by prions. Cattle acquire BSE mainly through eating the offal of sheep infected by scrapie—a feeding practice that has since been discontinued. Initially, Prusiner's research went largely unnoticed, but in 1997 he won the Nobel Prize in Medicine for his exemplary work.

The "Mad Cow scare" took place in the mid-1990s when over a dozen people were diagnosed with a new form of Creutzfeldt-Jakob Disease (CJD), another related prion-caused illness that devastates the brain. Strong circumstantial evidence indicated patients were infected by consuming meat from BSE-plagued cattle. (Some scientists feel, however, that proof of this connection has yet to be established.) A massive panic broke out in Britain, in which many who had recently eaten hamburgers and other forms of beef feared they would soon go mad and then die from a horrible illness. Some say the pressure to take action placed on the Conservative government that ruled at the time was one of the principal factors in bringing down that regime and replacing it with Labour. Eventually over a million cattle were destroyed in order to make sure BSE did not harm the British beef industry.

The lesson we might learn from the story of the discovery of prion infections is that generations of medical advances have yet to illuminate the darkest corners of the world of disease. Many mysteries remain about how deadly microbes carry out their sinister roles. New revelations about the generation, spread, and treatment of illness occur with remarkable frequency. For example, biologists used to think each disease is caused by a single agent. Now many believe viruses and other microorganisms can work in tandem. No one can truly anticipate what new ailments will challenge the medical teams of the future.

Imagine if a new airborne virus appeared—as contagious as the common cold, but as lethal as AIDS or Ebola. Within weeks, innumerable individuals would become infected. Ordinary casual contact, such as shaking hands, sharing of drinking glasses, or being exposed to sneezing, would in many cases become a death warrant. Over time, only those whose genetic make-up offered them strong resistance to the virus would avoid the consequences of exposure. If no cure could be found, myriads of afflicted people would perish. The loss of countless individuals—who once enriched the world with their vital talents and creative energies—would be unbearable.

Considering we have a difficult enough time dealing with the diseases which arise from nature, many of us would find it hard to fathom the possibility of human technology applied to creating especially potent microbes. Yet the history of our inhumanity against each other suggests evil minds may indeed be trying to perfect the ultimate biological weapon: an incurable, infectious illness with which to inflict their enemies. As long as nations misuse scientific research for military purposes, the threat of a biological apocalypse should certainly not be underestimated.

Garrett warns in her treatise about the growing threat of global biological warfare. She points out that although some 125 nations have signed the Bioweapons treaty, banning germ warfare, nothing insures it will be effective. Only the goodwill of nations, a rare commodity in times of strife, will guarantee that no such instrument be used.

In the recent film by Terry Gilliam, *12 Monkeys*, a deranged terrorist steals vials of an extremely potent, genetically-altered virus. He then flies around the world, releasing the virus in numerous major cities, and causing the deaths of billions of innocents. Only the handful that have developed some immunity to the illness survive. Then, however, the virus mutates into myriad forms and the remaining citizenry are forced underground into vast shelters. Their lives become sterile and meaningless, in a futile search for a cure for the disease.

Could a new, incredibly virulent disease wipe out all or most of humankind? One would like to think that modern medicine has advanced to the point in which such a nightmare scenario would be unthinkable. Unfortunately, medical science still has far to go before the possibility of biological armageddon becomes a matter of history.

In *12 Monkeys*, and many other fictional depictions of apocalypse, prophetic doomsday cults anticipate the end of the world with rash, often violent, behavior. The real-life versions of these groups have trumpeted their sirens for quite some time, warning in menacing tones of imminent global disaster, while piling up frightened devotees desperate to avoid armageddon. If a plague or other natural calamity occurs, they are typically among the first to exploit it as a tool for recruitment. Borrowing from biblical imagery, as well as from the mystical writings of medieval visionaries such as Joachim of Fiore, they often channel natural apprehensions about the end of the world into singular devotion toward a charismatic leader. Throughout the modern era, from William Miller in the nineteenth century, to Jim Jones, David Koresh, and Marshall Applewhite in the twentieth, impassioned forecasters of doom have attracted millions of eager adherents. The messages of these self-proclaimed prophets have resonated with particular strength in America, a land where religious liberty is considered sacrosanct.

CHAPTER THREE

THE WAR FOR ETERNITY

MODERN APOCALYPTIC NOTIONS

Red Alert! Hale–Bopp Brings Closure to Heaven's Gate.
Whether Hale–Bopp has a companion or not is irrelevant
from our perspective. However its arrival is joyously very
significant to us at 'Heaven's Gate' . . . Hale–Bopp is the
'marker' we've been waiting for—the time for the arrival of
the spacecraft from the Level Above Human to take us home
to 'Their World' in the literal Heavens.

—From the Final Web Site Message of Heaven's Gate

America's Sect Appeal

The United States, a nation with ample space, abundant resources, and proud traditions of rugged individualism and independence, has long been fertile ground for sectarian movements of all kinds. In a blending that no doubt confuses the rest of the world, a die-hard anarchistic spirit pervades the country with the most powerful government on earth. With no established religion, you can practice any one of many hundreds of different faiths or make up your own as long as you do not interfere with the workings of the civil order.

Many nations around the world—certain European countries, for example—possess a broad diversity of political beliefs, but harbor a bland uniformity of religious tenets. Not so in the United States, where—with politics tending to resemble neutral wallpaper —religion provides a brilliant multihued tapestry of divergent thoughts. (A few periods of intense political controversy, such as the 1930s, during the Depression, and the 1960s, during the civil rights movement and the Vietnam war, provide notable exceptions to this rule.) Unlike life in Europe, radical private faith generally forms a more important element of the American scene than does revolutionary public activism.

It is no wonder that the United States has served as a center for millenarian beliefs unequalled since the Middle Ages in Europe. With all faiths welcomed, and none seeming particularly odd or out of place, apocalyptic sectarians have historically sought a safe haven on the unassuming American soil. Indeed, well before the Statue of Liberty wielded her welcoming torch, several such groups discovered the American "religious experiment" to be a welcome change from stodgy, and often oppressive, European tradition. No holy inquisition, nor official church hierarchies, tormented free-thinkers once they crossed the Atlantic. The flagellants probably would have been welcome in the United States, along with other radical prophetic faiths, assuming they toned down their bullying, kept mainly to their own communities, and did not disrupt commerce. Often, not only did European millenarian groups take root in the American loam, they engendered native creeds from their far spreading saplings, until a veritable forest of radical religious culture was born.

Almost two years before the Declaration of Independence was signed, Mother Ann Lee, spiritual leader of The United Society of Believers in Christ's Second Appearance, better known as the Shakers, along with a small group of followers, first set foot on American shores. Their August 1774 arrival in New York Harbor marked the start of what was to be a blossoming of millenarian ideas in the United States.

The Shakers began in northern England as a Quaker offshoot. They were particularly interested in expunging humanity's sins in what they saw as the last days before the apocalypse. Taking the iconoclastic position that cohabitation of the sexes, even in matrimony, is one of the principal sources of sin, they condemned the Church of England for its support of marriage. Another sin—greed—

they thought could be abolished only through communal sharing of property. Christian purity, in their opinion, demands poverty. Believing that the end of the world was drawing closer and closer, Mother Ann and her fellow believers proclaimed their convictions increasingly more openly and strongly as time went on.

As the Shakers' critiques of the Anglicans grew more and more threatening, church officials decided to take action. Mother Ann was jailed for fourteen days on charges of blasphemy. Rather than diminish her influence, her brief incarceration served to install her as a martyr. By the time she left prison, she was venerated like a saint by her followers, many of whom began to consider her the female personification of Christ himself.

Escaping the possibility of further persecution, the Shakers relocated to New York and New England shortly thereafter. Free to exercise their beliefs, they established a number of new communities, where they practiced celibate, communal living and dedication to an unassuming, hard-working existence. The lifestyle they assumed was in accordance with their developing notion that the Second Coming had already occurred in the person of Mother Ann. They proclaimed the Millennium had begun in 1770, the year when Ann assumed her place as the female Christ, and that consequently one had to live one's life according to the divine principles she prescribed.

The sexual passion the Shakers forswore became channeled into their craftsmanship, as well as their frenetic forms of worship. Hence "Shakers" meant shaking Quakers. Shaker arts and crafts, especially Shaker furniture, is renowned for its elegant simplicity. Its sturdiness reflects the time and effort the devotees spent on its construction. What energy they had left over exploded into the holy anarchy of their services. Valentine Rathbun's personal account of a Shaker meeting in the eighteenth century describes this mayhem:

> When they meet together for their worship, they fall a groaning and a trembling, and every one acts alone for himself; one will fall prostrate on the floor, another on his knees, and his head in his hands; another will be muttering over inarticulate sounds, which neither they or any body else understand. Some will be singing, each one his own

tune . . . others will be agonizing, as though they were in
great pain; others jumping up and down; others fluttering
over somebody, and talking to them; others will be shooing
and hissing evil spirits out of the house, till the different
tunes, groaning, jumping, dancing, drumming, laughing,
talking and fluttering, shooing and hissing, makes a perfect
bedlam; this they call the worship of God.[1]

The arrival and establishment of the Shakers in America formed
just the start of an era of religious revival known as the Great Awak-
ening. For many enthusiastic Christians, the New World seemed the
perfect place to establish God's millennial kingdom—the holy city at
world's end as foretold in the book of Revelation. In the early nineteenth century, groups as disparate as the Church of Jesus Christ of
Latter-Day Saints (Mormons) and the Rappites (a German group
practicing communal living in Ohio and Indiana) attracted scores of
devout participants eager to live in the pioneering manner of early
Christians, at least as they perceived that lifestyle to be. While the
Shakers preached sexual abstinence formed the portal to God's community of grace, John Humphrey Noyes, founder of the Perfectionist
movement in Oneida, New York, advocated the diametric opposite
practice called complex marriage: complete sexual freedom within a
religious community. The Perfectionists' principle of the communal
sharing of one's sexual partners brought their community a measure
of unwanted publicity comparable to that experienced by the Shakers, albeit for fundamentally different reasons. Noyes rebutted criti-
cism of his "libertine" community at Oneida by arguing that the
abolition of monogamy, rather than being the relaxation of virtue,
constituted instead a supreme example of Christian munificence in
the millennial age.

While the Shakers, Perfectionists, and others joyously celebrated
what they saw as the final historical era by establishing new Christian communities, groups such as the Millerites took a sterner line.
Revelation's message to them was that humankind's doom approached shortly, and that the devout must quickly prepare. The imminent end of the world, they argued, should strike fear in one's
heart, not excitement, and be a cause for repentance, not ardor.

Great Disappointment

William Miller was born in 1782 to a improverished family in Pittsfield, Massachusetts. He is the founder of the Millerites (also known as the Adventists). While a youth, he was not particularly religious. Believing that God exists, but not in the same way that biblical accounts depict Him to be, he considered himself a deist. The Bible, he thought, contained too many contradictions, reversals, and inconsistencies to be of divine origin. Religion, as he first saw it, was an artifice, designed to lure men into church, and then onto the battlefield, defending so-called holy causes. As he wrote, "While I was a Deist, I believed in a God, but I could not, as I thought, believe that the Bible was the word of God."[2]

Though he did not believe that religious zeal was proper motivation for fighting, he thought that patriotism was such a reason. Transfixed with love of country, he entered the Vermont militia to fight in the War of 1812 against England. He fought long and hard, witnessing American victories that appeared to defy all odds. Miller became convinced that a Creator's hand must have contributed to these improbable gains. He abandoned his deism, and began to seek faith in a living, active God.

After constant exposure to the horrors of battle and the chilling air of senseless death, Miller left the military service an emotional wreck. In his time of need, he put his skepticism aside and turned to the Bible for comfort. He attended Baptist services and immersed himself in the Christian faith he had long shunned.

Like Augustine long before him, Miller brought a newcomer's fresh gaze to critical examinations of the Bible. His heart sought the answers to the questions his mind posed years before. He wanted so much to believe the Bible was God's word that he spent a considerable amount of time developing what he saw was reasonable explanations for the inconsistences within it—those issues that he once dismissed as intractable dilemmas.

Much to his delight, Miller found the evidence that he sought in the Bible to bolster his nascent faith. After two years of analyzing each of its verses, he concluded that each biblical prophesy ever made has eventually come true. Moreover, nothing appears in the

Bible by chance. Every comment and passage has an instructive or predictive purpose, ordained by God, he believed. All statements and references should be taken as literal, unless they are obvious metaphors. In short, he interpreted the Bible as a perfect guide to human life and human history, containing all possible truths and revealing all twists and turns of destiny. One must simply have faith that its prognostications will be eventually be fulfilled, it they haven't been already.

Miller's biblical analysis revealed a holy truth that seemed to call for urgent action; within decades, he determined, the world would be coming to an end. By painstakingly resolving and adding up chronologies in the Bible, he had become convinced that the last days of humanity, culminating in the Second Coming of Christ, would take place sometime in 1843. A detailed series of calculations, based on prophetic statements in the books of Daniel, Matthew, Revelation, and others, had led him to his shattering conclusion.

For years, Miller did little to advertise his results. By that time he was in his forties and fifties, and he considered himself too weathered to engage in the fiery rhetoric of an evangelistic preacher. He felt he was only "a poor feeble creature"[3] and an "old farmer,"[4] too pathetic to be taken seriously. Still, he wouldn't hide his beliefs and would announce them whenever he felt comfortable enough with the forum—for example, in the presence of clergy.

Every prophet can use a public relations person from time to time. Miller found his in Joshua V. Himes, an enthusiastic young minister and activist who was always looking around for a good cause with which to involve himself. Himes had applied his formidable energy to the temperance, women's rights, peace, and antislavery movements well before he had met Miller. In 1839, when he first encountered Miller at a Christian religious conference, Himes was struck by the prophet's exuberant spirit, which reminded him of his own. Though he wasn't entirely convinced at first that doomsday was imminent, Himes volunteered to spread Miller's message, believing that it would serve as a call for world peace and cooperation.

Himes founded the first Millerite newspaper, called *Signs of the Times*, and subsequently became its publisher and editor. The paper brought Miller's crusade, which had come to be known as Adventism, substantially greater recognition than it had ever received. In short order, Himes began several other periodicals to further spread the word, with names such as *Advent Witness, Midnight Cry, Trumpet Warning,* and *Voice of Alarm.* Following the demeanor of the times, these publications relied on sober, well-constructed reasoning, rather than superficial emotionality, to announce to the public that the world would end soon.

Miller's message, as conveyed by Himes, quickly began to attract a mass audience. Tens of thousands of devoted followers crammed into Millerite meetings, read its publications, and braced themselves for doomsday. Miller was invited to speak at numerous public meetings, church groups, and conferences. However, ill health precluded him from attending most of the events on his hectic schedule. Himes, who dropped most of his other activities as he became increasingly convinced of Miller's apocalyptic forecasts, found himself a frequent substitute for his sickly mentor.

In January 1843, the supposed start of the year of doom, public pressure forced Miller to be more specific about his time frame. He announced that apocalypse would play out according to a chronology determined by the Hebrew calendar, rather than the Gregorian. His calculations suggested that the apocalyptic era had not yet started, but rather would begin in several months. Pressed to choose a precise date, he refused, and remarked instead that the end would come sometime between March 21, 1843 and March 21, 1844. He declared that the last possible time for the Lord's Second Coming would be Spring 1844.

Ominously, in late February 1843 a comet appeared in the sky, brighter, at times, than any seen in the previous seven centuries. Known as the Great Comet of 1843, it blazed so strongly that it could be seen for many days in broad daylight. Its vaporous tail, formed by sublimation as it neared the Sun, stretched out a record length of more than two hundred million miles. This enormous trail of gas was long enough (assuming it was oriented in the correct direction) to span the orbits of four planets—from torrid Mercury out to frigid Mars—all at the same time.

The presence of such a blazing dagger aimed at the heart of the Solar System was the cause of great consternation. The falling star foretold in Revelation seemed to have finally arrived. Filled with anxiety about this apparent portent of ill fate, thousands of new disciples joined the ranks of the Millerites, hoping that God would be merciful to them on judgment day.

The Millerites waited and waited, praying for a just and compassionate conclusion to human history. When March 1844 came and passed, and Christ's chariot did not streak through the heavens, many became disillusioned. Miller, quite disappointed, apologized to his supporters, especially to Himes, for his apparent miscalculation.

Although some followers left, Miller's church regrouped, adjusting their eschatological calendar somewhat to accommodate the reality that the world was still around. A revised biblical interpretation, putting off doomsday just a bit, gradually became adopted by the organization. Proposed by Adventist preacher Samuel Snow, the new chronology supported an end date of October 22, 1844. Choosing a firm date helped to refocus the movement, further boosting its membership.

Drama intensified once more as the chill of Autumn arrived. In the October 16, 1844, edition of *The Advent Herald and Signs of the Times Reporter*, Himes announced that he had published the last issue before judgment day and was discontinuing the paper. Even Miller, who had been wary at first of accepting Snow's revised forecast, braced himself for Christ's return.

By late 1844, Adventism had won so many loyal adherents that when the prophesied date passed and the Great Disappointment (as it came to be called) began, many remained faithful to the cause. Rather than giving up their dreams of a new age under Christ, they concocted explanations for what had happened, or, rather, not happened. Perhaps God was testing their faith by offering false hope, some surmised, in a manner clearly at odds with Miller's doctrine that the Bible offered simple, honest predictions. Others, including Miller and Himes themselves, suggested eschewing exact doomsday prognostications, but instead preparing to accept Christ at any time.

As time went on the Millerite movement splintered into a number of smaller groups. A band of several hundred Adventists joined newspaper editor Enoch Jacobs in converting to Shakerism. They performed this leap because, as Georgia Tech historian Lawrence Foster reports:

> The final failure of the Millerite prediction of the specific date for Christ's literal Second Coming was emotionally shattering to many in the movement. Shock, grief, and perplexity were widespread. Derided by outsiders and unsure themselves of what went wrong, Second Advent believers desperately sought to salvage something out of the commitment which they had so sincerely devoted to the cause.[5]

A large percentage of those former Millerites who joined the Shakers left after only a few years. Though the Shakers offered enthusiastic devotion to the millenarian cause, their ascetic lifestyle was difficult to accept. Rather than commit to a lifetime of celibacy, many quit the group and ended up in more mainstream churches.

Another segment of former Millerites became interested in the theological ideas of Ellen G. White, founder of what was to become the Seventh-Day Adventists. She argued that October 22, 1844, was misinterpreted to be doomsday. According to a revelation she had, this date constituted the day in which Christ entered a heavenly sanctuary to prepare for his earthly arrival. Her group thereby managed to avoid the pratfalls of setting a fixed time for the apocalypse, while keeping the essence of the Adventist spirit.

Miller died on December 20, 1849. Inscribed on his tombstone is the fitting message: "At the appointed time the end shall be . . . But go thy way till the end be, for thou shall rest, and stand in thy lot at the end of days."[6]

Views from the Watchtower

Less than a decade after the Great Disappointment, and little over two years after the death of Miller, a child was born who would grow up to lead a millenarian movement much longer lasting than the Millerites. Charles Taze Russell, founder of the Bible Students, the predecessor of the Jehovah's Witnesses, arrived in the earthly realm on February 16, 1852, in Allegheny City, Pennsylvania, a suburb of Pittsburgh. His mother died when he was only nine, so he was raised by his father, a businessman. Wasting no time to educate his son in the family trade, Russell's father put him to work as a partner when he was eleven. By the time he was fourteen, Russell had dropped out of school and was comanaging a chain of retail stores in and around Pittsburgh.

As a boy, Russell had a great love for the Bible and was full of devotion to God. That faith was tested, however, when as a teenager he learned of the standard Christian belief in heaven and hell. Although he supported the idea of heavenly reward, he was mortified by the notion of eternal punishment. How could a just and loving God, he wondered, condemn a sentient being to fiery damnation?

Russell's doubts were allayed and his enthusiasm restored when in 1870 he attended an Adventist religious service. He found in the remnants of the Millerite movement much of what he had hoped for in religion. He sympathized with their strict interpretation of the Bible and their portrait of God as a compassionate figure, to be worshipped and loved, rather than feared. Also, he supported the Adventist belief that the world order was corrupt and deserved to be overthrown in a moral and spiritual revival. Finally, he wholeheartedly agreed with their conviction that the time of Armageddon was close at hand, when Christ would return to lead the faithful.

Inspired by his encounter, Russell formed a Bible study group, called the Bible Students, for whom he served as pastor. Due to his charismatic leadership and savvy organizational skills, the society expanded its membership and started a popular newsletter. Russell wrote numerous books and articles to promote his group—even producing some of the first motion pictures to advertise its principles. Some of his books included *Food for Thinking Christians* and the six volume series *Millennial Dawn*, which sold over fifteen million copies.

Like the Shakers and the Millerites, the Bible Students decried society's vices and urged its members to live lives as pure and Christlike as possible in anticipation of humanity's final days. Russell himself, though married, shared with his wife a strict vow of celibacy. Honest hard work formed an important part of their creed, and slothful behavior was condemned. This emphasis on laborious effort particularly extended to the group's outreach and funding activities. The church supported itself primarily through the painstaking collection of voluntary contributions, typically requested after distribution of its tracts.

In 1880, Russell established his personal theological vision with a series of forecasts about the end of the world. He meticulously read the scriptures to establish a timeline of past and future history, then predicted that 1914 would prove to be a year of monumental importance: the start of Christ's millennial kingdom on Earth.

Russell's prophecy was gleaned by supposing that exactly six thousand years would lapse between the time of Adam and the last days of mundane existence. This figure is the same derived by the medieval millenarians from Peter's biblical statement that one of God's days represents one thousand years of human history. Since it took six divine days for God to create the world, it similarly will take humankind six thousand years to fulfill their historical purpose. Numerous theological sources, in Judaism and other ancient religions as well as in Christianity, refer to this particular timeline. After this ordained interval is over, the next thousand years will constitute the seventh or "Sabbath" millennium, a time of peace and harmony.

According to Russell's timeline, the apocalyptic clock started ticking in A.D. 1874, six millennia after the time of Adam. That year, Christ secretly began to set up his millennial kingdom—a process, Russell predicted, that would take decades to complete. Once Christ's reign was fully established, the true age of peace would begin.

Like Miller and other apocalyptic prognosticators, Russell calculated the year that Adam first received the breath of life by adding up the times of various biblical chronicles and then working backward from the present. He placed Adam's creation at 4128 B.C., and supposed that he and Eve spent two years in the Garden of Eden. By adding six thousand years to this date, he produced his A.D. 1874 prediction. Other doomsday forecasters have placed the time of Adam centuries earlier or later—arriving at significantly different timelines. These variations stem from the fact that many epochs and events in the Bible have unspecified time frames. For example, nowhere is it written how much time passed between Adam's creation and his expulsion from the Garden. One might only venture a guess, as did Russell.

Unlike the Millerites, Russell was careful not to specify exactly when the world would literally be destroyed. Rather, he predicted that 1914 would represent the end of the era called the "Time of the Gentiles," and the beginning of the time in which God's kingdom would be fully established in heaven and earth. (It began to be set up in 1874, when he thought Christ secretly returned.) Between the time of his prediction and sometime around 1914, doomsday prophecies forecast in the Bible would be fulfilled, he argued.

The coming of the new age would prove turbulent, Russell and his followers warned. As foretold in Revelation, it would begin as a dark time of wars, earthquakes, floods, fires, and pestilence. Ultimately, earthly civilization itself would be destroyed, and replaced with a new order ruled from heaven by Christ and his select disciples. The faithful and pure would hold exalted positions in the kingdom to come.

The year 1914 arrived with the coming of the First World War, one of the bloodiest conflicts in history. Russell viewed the conflagration as one of the biblical tribulations that would herald the world's impending destruction. The war, in his mind, fulfilled the predictions that he made. Nevertheless, critics of Russell's movement argued that the essence of his prophecies—events associated with the arrival of Christ's kingdom—had failed to transpire.

Russell died in 1916, and was replaced as leader of the Bible Students by Joseph Franklyn Rutherford. Rutherford, a judge from Missouri, proved to have the power of persuasion, as well as the organizational skills, needed to refocus the group and bolster its membership. A tall man with a deep booming voice, he projected the voice of authority each time he lectured to the public. He was an extraordinarily prolific writer, publishing hundreds of books and pamphlets in the span of several decades.

Rutherford's well known rallying cry, "Millions now living will never die," drew tens of thousands of new followers to his cause. The catchy slogan, introduced in 1918, formed the title of lectures, workshops, books, and other literature. It summed up, in an easy to fathom manner, his theory that the world would end within a single generation.

Rutherford's theory attempted to explain why most of Russell's apocalyptic forecasts had yet to be fulfilled. Rutherford argued that though Russell specified the year—1914—when the transition from the earthly to heavenly kingdom would commence, he did not predict exactly when that process would be completed. Extending the prophecy of his predecessor, Rutherford preached the conclusion of human history would take place within decades. He emphasized again and again that members of the generation alive in 1914 would ultimately witness the end of days. This enigmatic prediction could, in theory, refer to anytime from 1918, when it was issued, until the early twenty-first century, when the last of those alive during World War I presumably will have passed away.

Today, the successor group to the Bible Students, the Jehovah's Witnesses, continues to thrive and attract enthusiastic new membership. No longer issuing specific timetables for when world history will cease (the last such prediction was made for 1975), its current leadership urges, instead, an attitude of watchful waiting for an unknown end time. This perspective is epitomized by the famous watchtower symbol that has come to represent the organization.

Tragedy in Jonestown

Millenarian movements from the nineteenth and early twentieth centuries generally lent their support to pacifism. Honoring Christ's dictum, "blessed are the peacemakers for they shall be called the children of God," they usually urged their membership not to participate in military conflict. Shakers would be shocked, for example, to see any members of their group bearing weaponry. Those who insisted upon doing so would be issued stern warnings, and then expelled if need be.

Over the years, many thousands of Jehovah's Witnesses around the world have valiantly defied governmental authorities and refused participation in the armed forces. Their conscientious objection to war often earns them lengthy prison terms under harsh conditions—or sometimes death.

Even societies with high degrees of religious tolerance—but afraid of mass resistance to conscription—have punished those who morally object to warfare. Rutherford, for example, along with many of his supporters, served jail time in the U.S. for his opposition to World War I.

Less tolerant societies have taken approaches that are far more severe. During World War II, numerous Witnesses were persecuted and killed by the Nazi regime in Europe. Still, in spite of the ridicule and hardships they face, church members argue vehemently that God's mandate for world peace, rather than civil law, must be obeyed as their highest authority.

Sadly, many apocalyptic movements founded during the late twentieth century have veered away from the tradition of peaceful preaching. Suspicious of governmental officials and other perceived outsiders, a number of groups have sermonized on the necessity (or at least the inevitability) of violent confrontation with the state. Often these suspicions have led to arms buildups, which have, in turn, precipitated either armed conflicts, mass suicides, or both. After such a disaster occurs—most famously in the 1978 Jonestown and 1993 Waco tragedies—some analysts blame the movement for what happened, while others criticize the government. In many cases, however, a tragic series of misunderstandings and miscalculations leads both the group and the authorities down the regrettable path to a showdown.

There are many reasons why modern sectarian religious groups attract so many zealous adherents. Family connections, once iron clad, now are generally much looser. Many lives contain emotional voids that membership in an alternative religion serves to fill, at least momentarily. Established religious faiths, traditionally a source of comfort for many worshipers, do not satiate all appetites in today's dynamic world. Those who yearn for action, rather than just contemplation, tend to seek out radical spiritual movements that strive for a new order, not just to preserve the status quo.

These considerations, however, do not wholly explain why some contemporary splinter groups are more prone to violence than earlier such societies. Why are today's sects more likely than yesterday's to collect armaments, poisonous substances, and other deadly materials, and then use them against themselves or others?

In my opinion, two principal factors have precipitated the current epidemic of violent confrontations and mass suicides. The first reason is the manifest increase of societal violence in general, particularly in the United States where weapons are readily bought, sold, and stockpiled. Despite new regulations, purchasing guns and ammunition has remained shockingly simple in many states. In some places, owning a firearm is not just considered a sacred right, it is also viewed as a sign of independence, a token of maturity, a commander of respect, a symbol of (male) virility, and an icon of freedom.

A popular slogan often spouted by avid gun owners is that guns don't kill; people kill. Its obvious implication is that guns cause harm only when in the wrong hands. Though the expression is patently misleading—countless deaths caused by firearms occur because of accidents, mistakes, and hasty reactions during quarrels—it does contain a grain of truth. Gun violence in the United States is not only due to the abundance of weapons, but also stems from a culture that emphasizes dogged competition.

Many Americans are driven by the insatiable quest to "get ahead" of others by acquiring as many luxury items as possible. A large number of impoverished and poorly paid individuals, however, are financially unable to join this race to the top. Because they lack the money to live "the good life" that defines American culture, many naturally feel bitter and disempowered. Some turn to violence as a means of venting their anger and crime as a way of gaining what would be impossible for them to earn in the workplace. Thus, in a society in which material possessions define the person—and yet many lack basic necessities—is it any wonder that enraged, violent behavior is so common?

The second major cause of the wave of apocalyptic confrontation that especially challenges American society is the reduction of privacy with the vanishing of frontiers. Paradoxically, though individual liberties form a supreme feature of American life, that freedom has become increasingly challenged by a growing network of transportation, communications, data transmission, and processing, coupled with a steady diminution of available space.

In the past, groups that considered themselves radically different from the majority could readily find solitude by separating themselves physically as well as philosophically from the pack. If neighbors complained too much, or if they subjected the organization to too many intrusive notions that would undercut its autonomy—threatening its independence and purity—then it could just pick up and leave. Enough land was available to accommodate all. Transport was slow and communication haphazard, allowing for easy isolation whenever desired.

Today an intricate spider web of highways and other well paved roads link all but the smallest American communities, making complete seclusion much harder to achieve. Aside from relocating itself in one of the few truly remote regions left, such as northern Alaska, the principal way a community might find solitude is to gate itself in, an act some might misinterpret as being hostile to outsiders. Ironically this perception of hostility might lead to greater scrutiny of the group by neighbors, complaints to the government, and, ultimately, confrontation.

Moreover, because of radio, television, and the Internet, contemporary alternative communities cannot rely on physical isolation to foster their separate and unique programs. Through audiovisual and computer-based media their members are constantly exposed to the outside (gentile) world. Conversely, the greater community, if it wishes, can learn about their activities through revealing newspaper and television investigations.

Increased risk of exposure to and contamination from external influences forces each sectarian group to make the difficult choice of either greater accommodation and blending in with the outside world, or else strict authoritarian control of its boundaries—its physical perimeters as well as its media access. Communities choosing the former path take the chance their message will eventually become so watered down that their membership can no longer be distinguished in its lifestyle from other Americans. Those who embark on the latter route, on the other hand, find themselves literally forced to barricade themselves into armed fortresses.

To counteract media images and other information from beyond their enclaves, they sometimes end up relying on subtle or overt forms of intense psychological persuasion (called "brainwashing" or "mind control" by critics, but referred to as "education" or "enlightenment" by supporters). These methods might include sleep deprivation, sexual humiliation and control, control over bodily functions (denial of bathroom privileges), continuous exposure to repetitive audio and video tapes, and malnutrition. Research has shown that hungry, sleep-deprived people are more likely to be a placid, eager flock, than those whose basic needs are met.

A common feature of radical sectarian religious groups is the presence of charismatic leadership, often serving emotionally as surrogate parents. Feeling isolated, and perhaps misunderstood by their real friends and families, group members rally around these central figures, clinging to their leaders' comforting authority. Presiding over enclaves isolated behind sturdy barricades, these leaders can create their own fiefdoms, dominating all facets of life, and furthering the sense of abhorrence of the outside world. Seclusion combined with mistrust can escalate into a paranoid sense of being under siege, finally resulting in armed struggle.

Consider, for instance, the tragic story of the Peoples Temple, a California-based group led in the 1970s by the Reverend Jim Jones. In one of the greatest disasters of the late twentieth century, 913 members of the cult, including Jones, died from poisoning (voluntarily self-induced, as well as coerced) or gunshot. This mass suicide/homicide shattered forever the myth that religious communal living always strives toward peace and contentment. A new image of the commune as armed camp became indelibly impressed upon the public's perception.

Originally headquartered in San Francisco, and dedicated to eliminating racism, the Peoples Temple moved to Guyana to set up its own "model community," called Jonestown. There, Jones established his fiendish order—authoritarian rule through psychological manipulation and physical terror. Humiliation, denigration, and punishment were all instruments in Jones' tool kit as he exerted almost godlike control over his followers. Acting as members' friend, foe, father, and lover, he would exploit their emotional weaknesses in order to satisfy his desire for total domination.

As Keith Harrary, a prominent psychologist who counseled former cult members and their families, reports:

> There was a deliberate malevolence about the way Jones treated the members of his cult that went beyond mere per-

version. It was all about forcing members to experience themselves as vulgar and despicable people who could never return to a normal life outside of the group. It was about destroying any personal relationships that might come ahead of the relationship each individual member had with him. It was about terrorizing children and turning them against their parents. . . . In short, it was about mind control.[7]

Ultimately, Jones acquired a grotesque longing which led to the group's undoing: the power of granting life and death. Suicide drills, in which he would ask his followers to drink wine, and then tell them (falsely) that it had been poisoned, comprised a way in which he began to routinely test their loyalty. He warned anyone who considered escaping that his personal thugs (called angels) would kill them if they tried.

As news about the horrors of Jonestown became reported back to the United States, pressure mounted for a governmental investigation. Cult members' families insisted the government find out what was really happening. Learning about the imminent threat of an inquiry, Jones armed selective members of the group and braced them for battle.

Finally, California Congressman Leo Ryan and a select group of reporters flew to Guyana in an attempt to find out the truth about Jonestown. Ryan met with some of his former constituents, and offered them safe passage back to the States. Only a handful accepted his offer. On the way back to the airport, Jones had the group ambushed and murdered.

Fearing reprisal for his actions, Jones decided to hasten Armageddon. He immediately ordered his flock to commit mass suicide. The few survivors of the tragedy at Jonestown describe what then ensued as an incomprehensible spectacle of horror. Hundreds of members quaffed cyanide-laced fruit drinks that were handed out to them. Some drank willingly, while others were forced at gunpoint. Yet others, including Jones, were shot or stabbed by unknown assailants. Jonestown's reign of terror drew to a hasty, bloody conclusion—in a catastrophe worse than even the most cynical observer would have predicted.

Might hundreds of lives have been spared if Ryan hadn't visited? No one knows for certain. Some argue today that the tragedy at Jonestown constituted the inevitable outcome of the warped society Jones designed and led. Since Jones had already conducted suicide drills, mass annihilation was clearly in the cards. Others stress that compassionate dialogue with alternative religious groups might allay misunderstandings and reduce the chances of confrontation. If Jones hadn't felt so threatened, they assert, perhaps he would not have issued his murderous orders. Since Jones is not around to register what he perceived at the time, this controversy cannot be settled.

Ranch Apocalypse

The debate about government treatment of sectarian religious groups took a disastrous turn in 1993, when eighty-six members (this figure is disputed) of the Branch Davidian order of the Seven Seals died in a tragic fire at their complex, the Ranch Apocalypse near Waco, Texas. The conflagration began after the federal government tried to evict the group from their complex. With arguably greater justification than in the Jonestown situation, a number of vocal critics feel that if the government had been more careful, the Waco disaster might well have been avoided.

The Branch Davidian movement is a splinter group of the Davidians, who, in turn, trace their origins back to the Adventists. Victor Houteff, founder of the Davidians, emigrated to the United States from Bulgaria in 1907. After attending numerous meetings of the Seventh-Day Adventists (one of the successors to the Millerites), he decided in 1929 that he was destined to lead that church along the path of a new spiritual revival. After his efforts to steer the Seventh-Day Adventists along a more radical millenarian course were resisted, he elected to form his own alternative denomination.

In 1935, Houteff, along with two of his followers, bought 377 acres of land in Mt. Carmel, near Waco, Texas. There Houteff established a new religious community, based on his views of how Christians should live in the last days before Armageddon. Loosely borrowing from the ideas of William Miller and Ellen White, he taught that those expecting to participate in Christ's millennial kingdom must divorce themselves from worldly affairs, live simple, frugal lives, and study the Bible closely, particularly its prophetic elements.

Following their leader's dictums, most of those residing in Mt. Carmel took to farming as their principal means of support. Farming allowed the community self-sufficiency. Davidians were able to work, eat, and sleep in their community without having much contact with outsiders.

When Victor Houteff died in 1955, his wife Florence assumed leadership. She was challenged for that position by community member Ben Roden. When Roden failed in his leadership bid, he wrote a letter to the Davidians in which he called himself "Branch," after a passage in Zechariah, "a man whose name is the Branch shall build the temple of the Lord." Roden left Mt. Carmel, and formed his own group, the Branch Davidians.

As the new prophet of the Davidians at Mt. Carmel, Florence Houteff took a step beyond which Victor had been willing to take; she predicted when the world would end. The last hours of earthly life, she admonished, would occur on April 22, 1959—coinciding with the Jewish holiday of Passover. Apparently, she struck a chord with many believers. Hundreds left all of their possessions behind—frantically selling their houses and businesses—and joined her at Mt. Carmel.

When Florence Houteff's prophecy failed to transpire, another Great Disappointment ensued, similar to what happened after Miller's prediction proved unsuccessful. The bulk of those who had converged on Mt. Carmel went home gravely disappointed.

In 1962, her community in turmoil, Florence Houteff decided to dissolve the Davidian association, and move to California. A chaotic reign began in Mt. Carmel that was to last for several decades.

The Branch Davidians took over the Mt. Carmel complex, and reestablished a settlement there under the leadership of Ben Roden, along with his wife Lois and son George. The Rodens ran the settlement for more than a decade, promoting Ben's idea that practicing all of the biblical holidays (Passover, etc.) would hasten the time of the Second Coming.

When Ben died in 1977, Lois and George engaged in a power struggle for control of the group. Lois emerged as the new leader, and appointed young enthusiast Vernon Howell to be her successor.

Then, in 1985, Lois died. George Roden and Vernon Howell fought fiercely for leadership, even exchanging gunfire at one point. Finally, in 1988, Howell won the battle and asserted control of Mt. Carmel. He adopted the name "David Koresh," (in reference to King David, and, perhaps, to Persian King Cyrus) and began the arduous task of rebuilding a shattered community.

Superficially, Koresh seemed the Elvis of apocalyptic leaders. He sang songs, played the guitar, and could talk to practically anyone about practically anything. By all measures, he was charming, friendly, and attractive. His charismatic personality won the group dozens of converts and hundreds of thousands of dollars in donations. Yet his easy manner masked a stern and domineering side, an attitude that his detractors called sadistic. His advocacy of the wide use of corporal punishment of children eventually led to allegations of child abuse against him. He separated family members from one another (children from their fathers at birth, and from their mothers at age twelve), persuading all youngsters to think of him as their father, and all women whom he was interested in to embrace him as their lover. Because of his congeniality, and claims to biblical authority, these requests were readily accepted by the group.

Holding strong persuasive power over his flock, Koresh came to think of himself as the Lamb of God. He believed that he had discovered the secret code for opening the seven seals of the holy scroll of destiny described in Revelation. Once this parchment is unsealed, according to Scripture, the forces of darkness will be unleashed, and Doomsday will be close at hand. Consequently, Koresh felt that he alone was able to bring on the Apocalypse. He renamed Mt. Carmel, "Ranch Apocalypse," stockpiled it with weapons intended for the final battle against the "forces of evil" (namely, the civil authorities), and warned his followers that the end was near.

Over time, the federal Bureau of Alcohol, Tobacco and Firearms gathered evidence that Koresh's group had accumulated hundreds of machine guns, handguns, rifles, grenades, and other weapons, either bought or assembled from parts. Trying to seize this emerging arsenal, agents from the bureau attempted a surprise raid on the complex. Koresh caught wind of their plans, and ordered his supporters to prevent the agents from entering. A gun battle ensued that killed four of the agents, wounded fifteen more, and left an unknown number of Koresh's followers dead or injured.

After the firefight was over, pressure mounted for a resolution of what had become an explosive situation. Cult-watchers, fearing another Jonestown, warned that Koresh's group was unstable, and likely to commit mass suicide if provoked. Some experts (including United States Attorney General Janet Reno) argued that the children inside the complex were victims of abuse, and should be considered hostages. Meanwhile, the government raid provided further justification for Koresh to argue to his members that they were being unjustly persecuted.

A fifty-one-day standoff ensued, in which the Federal Bureau of Investigation tried to force Koresh and his followers from Mt. Carmel. Applying what they saw as psychological pressure on the group, they bombarded the complex with loud, droning music and called upon group members to surrender. Finally, when the FBI seemed to be at its wit's end, its agents used tanks to punch holes in the compound's walls, and then poured tear gas in the openings.

The incursion was met with a sudden blaze of fire, which grew into a veritable inferno. Apparently followers of Koresh sought to hasten the end time through a literal conflagration. Trapped inside, dozens of men, women, and children perished. The burnt body of Koresh was discovered among the victims.

Who should take the blame for the Waco catastrophe? Some pin it squarely on Koresh for his dominating attitude and violent behavior, others on the government for miscalculating Koresh's motives and reactions. Yet others suggest that the great availability of firearms in the United States provided the sure recipe for disaster.

Social scientists, who have developed models of cult mentalities, assert that a better understanding of Koresh's beliefs, including his assertion that he held the key to unlocking the seven seals, may have reduced the chances of violent confrontation with his group. Some argue that Koresh would have been more likely to negotiate with those who appreciated his principles and prophecies. However, since there is only so much one can glean at any given time about the mentality and potential for destruction of a secluded, inaccessible individual, it is unclear if a stronger attempt to understand Koresh would have significantly changed what ultimately took place.

All would agree, however, that the script that ended up being played out at Waco resulted in one of the worst possible outcomes imaginable. Let us hope that such an ill-fated series of events never happens again.

Souls to the Heavens

A curious characteristic of several alternative religious groups in the 1990s has been a coupling of traditional apocalyptic fervor with the notion held by UFO-devotees that extraterrestrials have contacted Earth. In this hybrid of beliefs, Christ's Second Coming will take place via an interstellar journey. When the world is fated to end, he will rescue deserving earth dwellers, and carry them on his space ship (or another transportation device) to safe haven on another planet. Meanwhile, the rest of us left behind on Earth will suffer through Doomsday.

Two recent organizations in particular, the Order of the Solar Temple and Heaven's Gate, have convinced many of their members that apocalypse was imminent, and that they stood to be saved by being brought to a new world. In a series of horrific incidents, followers of each group ritually killed themselves, believing that this opened the door to their being transported into space.

The Solar Temple (International Chivalric Order Solar Tradition) was founded by Belgian homeopathic expert Luc Jouret in 1977, as the successor to a group called Foundation Golden Way. Like Jones and Koresh, Jouret believed that the end of creation was imminent, and that his leadership provided the only means of salvation. He proclaimed that he alone could lead humanity's elite to a new life on a planet near Sirius. Preaching an amalgamation of ideas derived from the mystical Knights Templar movement (a medieval secret society), as well as from Revelation and astrology, Jouret established branches of his group in Switzerland, Quebec, and Australia, attracting hundreds of followers.

In October 1994, authorities were stunned to find waves of mass suicide and murder were spreading through villages in Switzerland and Quebec. That year, fifty-three people died, including Jouret, by means of immolation and other methods. The body count grew in

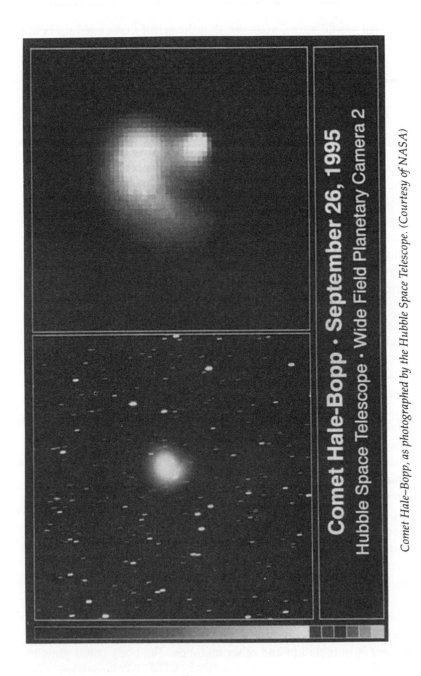

Comet Hale-Bopp · September 26, 1995
Hubble Space Telescope · Wide Field Planetary Camera 2

Comet Hale–Bopp, as photographed by the Hubble Space Telescope. (Courtesy of NASA)

1996, when thirteen adults and three children were found dead in a French forest. Still more deaths occurred in early 1997 in Quebec. Because of ceremonial robes, swords, and other paraphernalia found near the bodies, as well as suicide notes in some cases, experts assume these suicides and murders were attempts by Solar Temple members to make contact with the beyond, and travel to their aspired safe haven near Sirius.

When Comet Hale–Bopp was discovered to be approaching our region of space, experts on doomsday cults pointed out the prophetic significance of such bodies throughout history and predicted a significant possibility of disaster. Sadly enough they were right. The "omen" in a strange way turned out to be true.

In March 1997, the largest mass suicide of this century in the United States took place. The bodies of thirty-nine men and women, ritually draped with purple shrouds, were found lying on beds in a California mansion. Videotapes mailed out by members before their deaths confirmed they killed themselves for spiritual reasons. Their declared aim was to abandon their "earthly vessels" and travel on a spaceship docked behind Hale–Bopp.

The deceased had belonged to the high-tech Heaven's Gate apocalyptic movement, led by Marshall Appelwhite. Appelwhite, known to the group as "Do" (as in the musical notation, "do, re, mi"), preached that Revelation contained the prophecy that a UFO would rescue elite humans before Earth is someday destroyed. Followers of Do practiced celibacy and kept their heads and bodies shaven in an effort to keep pure and ready for the new world. Some of the men, for this purpose, even submitted to voluntary castration. They supported themselves by designing computer web pages. After their suicide, news media were startled to discover they had advertised their intentions to leave Earth on a specially designed website.

In the late twentieth century, an unprecedented wave of apocalyptic fervor has swept the world, particularly the United States. Though turning of the millennium is one of the obvious causes, other critical factors have triggered this boom. Among these are the varied new threats to earth's ecological balance. For the first time in history, humankind possesses the capability of causing its own demise through global ecological damage.

One of the poignant symbols of the past century is the specter of the mushroom cloud. In 1945, science proved capable of decimating entire cities through atomic explosions. In a literal flash, humanity was forced to come to grips with the realistic possibility of nuclear armageddon. Anxiety mounted as sabers rattled, tensions flared, bomb shelters were constructed, and drills coordinated. No wonder apocalyptic religious groups found themselves flooded with new recruits.

Today, with the Cold War over, the threat of nuclear warfare seems less substantial. However, weapons still exist, scattered around the globe, that if ever used could instantly decimate civilization, and perhaps even destroy our race. Barring the dismantling and destruction of these armaments, and the erasure from history of the knowledge to ever build them again, we must remain vigilant against their use. For if they are ever triggered, earth would become a cold, miserable, poisonous, and possibly lifeless relic of its former self.

The fiercesome potential for nuclear catastrophe is far from the only danger facing our planet's ecology. In recent decades excessive quantities of carbon dioxide, released into the air through industry and other human endeavors, have trapped the world's surface heat and caused average temperatures to rise. Scientists expect this climatic effect (known as global warming) will lead to a continuous rise in ocean levels, resulting in the flooding of many island and coastal communities.

In another prominent example of humankind endangering itself through pollution, scientists have found that artificial chemicals called chlorofluorocarbons (CFCs) are eating away at the ozone layer: the portion of the upper atmosphere that most effectively blocks harmful ultraviolet solar radiation. Much of this protective shield is destroyed already; even more is expected to be damaged in the near future. Researchers expect elevated levels of ultraviolet light due to the thinning of upper-atmospheric ozone will cause skin cancer and cataract rates to skyrocket in coming decades.

Today Earth provides a reasonably comfortable abode for the human race. Unfortunately, if the past is any guide, we cannot expect that current environmental conditions will continue forever. Even if the calamities produced through human folly can be avoided, nature will surely provide its own challenges—as it has many times already.

Tens of thousands of years ago, our planet experienced its most recent ice age: a prolonged period in which its average temperature was lower than today, and much of its northern hemisphere was covered by glaciers. Most geologists believe these harsh conditions will someday return. (Perhaps a period of sharp global cooling will follow the current age of global warming.) Though no one can predict exactly when the next ice age will commence, how long it will last, and by how much its effects will be reduced or superceded by other ecological factors (such as the warming influence of atmospheric carbon dioxide pollution, for example), almost certainly a future of momentous environmental challenges lies in store for our world.

PART II

ANGRY EARTH

WORLDLY APOCALYPSE

CHAPTER FOUR

THE SNOWS OF YOSEMITE

ICE AGE CYCLES

*Thus, by forces seemingly antagonistic and destructive, has
Mother Nature accomplished her beneficent designs—now
a flood of fire, now a flood of ice, now a flood of water; and at
length an outburst of organic life . . .*

—JOHN MUIR, *"The Mountains of California"*

Forever Winter

Yosemite National Park—as fans of the photographs of Ansel
Adams and the writings of John Muir would attest—is one of
the most beautiful scenic attractions in America any time of
year. In winter, its rocky formations are a jewel box: shimmering
white granite encased in glistening ice, sprinkled with majestic emer-
ald trees. Then, as the heat of spring and summer thaws out the
Sierra Mountains, melted snow drips down cascading waterfalls—
among the world's highest—into the natural splendor of Yosemite
Valley.

In the center of the park lies the unusual sight known as Half
Dome. Resembling a mountain broken in two, many geologists believe it is literally that. Tens of thousands of years ago, according to
theory, it was split in half by a great glacier. The icy river pulled off

the looser material, spreading it throughout Yosemite Valley, leaving behind the sturdier stuff that comprises the present formation.

Indeed, enormous rocks scattered among the evergreens of Yosemite stand as testimony to sweeping forces that reshaped the topography of the northern hemisphere. Isolated boulders, some as large as tanks, look strangely out of place among the rolling forested terrain. Appearing like solitary relics of a long gone monumental battle, the rocks, when chemically analyzed, show marked resemblance to geological material from more northerly regions. Because of this similarity, researchers believe the rocks traveled from more northern latitudes to their current positions. Swept to Yosemite by powerful natural forces, these one-time marauders from the north now stand like rusted hulks on an extinct battlefield.

Neither wind, nor water, nor any other ordinary type of erosion could have caused such cataclysmic shifts. Only the sheer power of flowing sheets of ice could have hoisted these rocks into place. Such unmistakable evidence has led the geological community to conclude that several successive eras of advancing and retreating ice took place in the past. The most recent ice age peaked roughly eighteen thousand years ago, long before California was first settled by native Americans.

Yosemite is not the only relic of the glacial ages of the past. Boulders now lying in Central Park, New York, were strewn there by mammoth ice flows from the north. The Great Lakes and the Finger Lakes were each etched out by glaciers and deposited with their melted waters. All around North America, Europe, and Asia lies evidence of the cataclysmic movement of materials relayed for thousands of years by frozen rivers.

Will it happen again? No one knows if and when another long icy cloak will blanket the north. The reason for these ice ages and other major climate changes is presently a matter of scientific debate. Several competing theories of why eras of glaciation and melting have rhythmically occurred continue to vie for preeminence.

Seasons of Ice

In the late nineteenth century, Scottish geologist James Croll advanced what has become the most widely accepted explanation of why the Earth passes through glacial (ice age) and interglacial (in between ice age) periods. Croll demonstrated that the Earth's orbit and tilt vary periodically, and that these variations significantly affect global climate.

Currently, Earth's axis tilts at an angle of 23.5 degrees away from its orbital plane. This inclination explains why days are shorter in winter and longer in summer. During the winter, the northern hemisphere tilts away from the Sun and receives fewer hours of sunlight; in summer, the opposite effect provides longer daylight hours. If the Earth's axis were perpendicular to its plane of orbit, days and nights would universally be equal in duration. If, in contrast, Earth were tilted along its orbital plane, like a spinning top that has fallen on its side, then half the world would experience twenty-four-hour days, while the other would experience twenty-four-hour nights.

Geological evidence indicates that Earth's tilt has varied over the ages from a minimum angle of 22.5 degrees to a maximum of 24 degrees, cycling between these extremes about once every forty-one thousand years. By affecting the average amount of sunlight received by the northern hemisphere during the winter and summer months, these variations have produced climate alterations. For example, winters in which days are slightly longer would be a bit more moderate than ones in which days are shorter, because, in the former case, the environment would be bathed in the Sun's direct energy for a greater interval each day.

Periodic alteration in the shape of Earth's orbit around the Sun forms another astronomical factor that may have affected terrestrial climate. Though Earth always follows an elliptical (oval) orbit around its parent star, throughout the ages its path has regularly undergone compression (to a near circle) and stretching (to a more oval shape). This feature, called orbital eccentricity, was shown by Croll to fluctuate about once every 100,000 years—possibly explaining similar patterns of glaciation.

Earth's orbital procession (turning) comprises yet another feature that may have induced global climatic transformations. As Earth's orbit slowly advances, the season when our planet lies closest to the Sun gradually shifts. Currently, Earth is nearest during the northern hemisphere's winter (when the North Pole points away from the Sun), and farthest away during its summer (when the

North Pole points toward the Sun). However, because of the procession of our planet's orbit, these conditions cycle with a periodicity of twenty-one thousand years, regularly affecting the amount and distribution of solar heat received on Earth.

It is unlikely these orbital factors could have affected average global temperatures by more than a few degrees. Yet as Serbian mathematician Milutin Milankovich demonstrated in the 1920s, small changes in how the world's solar energy is distributed could significantly influence its climate. He calculated that a gradual shift in the focus of the Sun's direct radiation from one region of our planet to another could conceivably set off an ice age.

Let us consider how a minor, but sustained, decrease in the average annual temperature of the northern hemisphere could eventually lead to a glacial era. Winter, spring, and summer would all be slightly cooler. Consequently, as time went on, at lower and lower latitudes, snow accumulated during winter would have a greater chance of melting later during spring, and perhaps even lasting until summer.

This effect would tend to build on itself, leading to thicker and thicker layers of snow and ice. A thin coating of snow covering the ground would keep it cool enough for additional layers to form more readily. Eventually, in once-moderate regions, a covering of packed snow would remain year round. Ultimately, these would become mammoth glaciers, cutting their way through rock and soil while pushing farther and farther south. Thus, in summary, as Milankovich suggested, small variations in the quantity and distribution of solar radiation received by Earth would become translated into epochs of glaciation.

Gradually the geological community has come to realize that the Croll–Milankovich model, though of great value, cannot represent the full picture of why ice ages occur. Fossil evidence indicates that global climate fluctuates between colder and warmer intervals far more frequently than astronomical patterns would indicate. Changes in Earth's tilt and orbit occur on too great a time scale to explain significant climatic changes that have taken place in centuries, even decades. For example, in a well-researched period 10,800 years ago, known as the "Younger Dryas," frigid conditions akin to an ice age swept quickly (as geological ages go) through Europe—arriving in less than a century and dissipating but six centuries later.[1]

Flash Frozen

The Younger Dryas represents one of the biggest puzzles in climatology. The abrupt changes in climate during that era cannot be explained by astronomical models. Since clearly the Earth could not have changed its tilt or orbit in the timespan of decades, the Younger Dryas phenomenon points to the existence of an amalgamation of factors that have an impact upon climate in a complex manner.

About eleven thousand years ago, the most recent ice age had just ended. Temperatures in North America and Europe were rising, and glaciers were in retreat. Flora and fauna native to more temperate climes were returning to what for tens of thousands of years had been bleak tundra. Beetles, for instance, moved into places such as England, displacing species better adapted to the cold.

Then bizarrely, in less than a century, the warming trend in Western Europe (but not in most of North America) completely turned around. It was as if someone grabbed an automobile stick shift and, without even braking, switched directly from drive into reverse. Species that did well in cold climates hastily returned to enjoy the more frigid environment, such as the beautiful Younger Dryas flower that gives that period its name.

The "mini ice age" lasted from about 10,800 years ago to 10,200 years ago. Then, at last, the last vestiges of the glacial era left Europe. Frigid plants and animals beat a retreat, temperate ones came back, and European weather became pretty much what it is today.

Why did Europe, and not America, experience such a double take? Scientists believe they know the answer. The currently favored explanation for the Younger Dryas period relates to the melting of the Canadian and European ice sheets at the close of the most recent ice age. Fresh meltwater from these mammoth glaciers poured relentlessly into the northern Atlantic below Greenland, changing its salt level for many centuries. Because fresh water is much lighter than salt water, it remained near the top of the ocean, forcing the extant saltier fluid to sink below it.

Another property of fresh water is that it freezes much easier than if it had salt. Therefore, the newly formed layer of meltwater blanketing the northern Atlantic had a much greater propensity to develop ice than did the earlier, saltier surface. Ice did form, blocking the ocean's ability to radiate heat—critical to the warming of the European coast—and pushing the more temperate Gulf Steam current much farther south for many years.

The moderate nature of European weather is critically dependent on the presence of the Gulf Stream. The Gulf Stream is an ocean circulation phenomenon that transports warm water from the tropical Caribbean sea and Gulf of Mexico to regions of the northern Atlantic. Near Europe, its heat gradually dissipates into the air, allowing places like Norway and Denmark to experience much milder winters than one would expect for their latitudes. The sea along much of the European coastline is like a colossal radiator, serving as a vast warm water heating system. After all, if one looks at the map, Scandinavia, Britain, Ireland, and the rest of northern Europe are each as far north as Canada. Yet none suffers through the harshness of a frigid Canadian January. Thank the Gulf Stream for that.

Whenever the Gulf Stream is diverted, Europe loses its cheap heat. It might bang on the pipes like an irate tenant, asking for its service to be turned back on, but it has no building superintendent that has the capacity to do so. Rather it must wait—for centuries or even millennia—until the Gulf Stream returns to its coast.

In the case of the Younger Dryas, it was six centuries until the waters of the northern Atlantic were sufficiently ice free so the Gulf Stream could resume its beneficent course. This retreat and readvancement of currents has been well documented by researchers William Ruddiman and Andrew McIntyre of Lamont–Doherty Geological Observatory. They used fossil plankton (simple, drifting forms of primitive plant life) in the sediments beneath the Atlantic to trace the history of warm and cold water masses. Icy ocean waters, they demonstrated, moved noticeably southward around the time of the start of the Younger Dryas, and northward again around the age of its conclusion.[2]

The story of the Younger Dryas demonstrates how global and regional climates are not just determined by how much heat Earth receives from the Sun due to its tilt and orbit. Rather, a host of internal factors play strong roles in influencing conditions. In this case, we see how ocean movement and composition feature prominently in setting the pace for local climate shifts, especially along coastlines. Another critical factor in the emerging picture of Earth's complex climate is the nature of its atmospheric content. As known for more than a century, the quantities of certain heat-trapping gases in the air, particularly carbon dioxide, have a crucial effect on Earth's climatic portrait.

Change in the Weather

A key to a deeper comprehension of what causes radical climate shifts was provided in 1896 by the renowned Swedish physical chemist Svante August Arrhenius. Arrhenius firmly believed that global changes in temperature stemmed from internal, as well as external causes. He developed a terrestrial interpretation of the ice ages—relating to how atmospheric gases trap and dispel heat—that supplemented the favored astronomical explanation.

In the Arrhenius model, the Earth's environment is like a house with many specially-designed windows. These windows are constructed so the homeowner can adjust how much heat (from sunlight) they let into the house, and how much energy they permit to escape. In January, for example, the windows are set to let in as much warmth from the Sun as possible, and to block heat from the hearth fires within from leaving. In July, on the other hand, the windows shade dwellers from the scorching heat outside, while permitting any radiation inside to flee and cool the house down.

Unfortunately, when Mr. and Mrs. Cryon, two homeowners, call up a home supplies distributor and order these miracle windows for their spacious villa in Edmonton, Canada, they receive defective versions with the controls strangely reversed. Through the long Canadian winter, they are appalled to discover their windows block off all incoming radiation from the Sun, while allowing all energy produced within to escape freely. Even at full blast, their furnace does not warm the air fast enough to compensate such a loss of heat. Temperatures drop to freezing, as the Cryons shiver and curse the day they purchased their new windows. In short order, their house resembles a large, tastefully-decorated icebox.

In similar manner to such hypothetical windows, Arrhenius discovered the chemical composition of the atmosphere, particularly its carbon dioxide content, regulates how much solar radiation reaches the surface of our planet, as well as how much of Earth's heat radiates back into space. Carbon dioxide has the property of allowing sunlight to stream toward Earth, but blocking surface heat from escaping. Whenever the quantity of carbon dioxide present in the atmosphere increases, a greater percentage of terrestrial surface energy becomes trapped, and Earth's average temperatures tend to increase. On the other hand, if the carbon dioxide atmospheric content drops, less surface heat is contained, and average temperatures plummet.

Arrhenius theorized that glacial periods correspond to eras of much lower than normal atmospheric carbon dioxide. Indeed, core samples taken from the ice cap in Antarctica demonstrate Arrhenius's postulate is correct. The frigid air of the ice age world contained significantly less carbon dioxide than at present.

Many climatologists believe internal oscillations of terrestrial conditions (land, sea, and air) contribute just as great an effect on the climate as do the external influence of astronomical rhythms. It is now well known that continents rise, fall, and drift, depending on geological circumstances. These land movements cause changes in air currents and ocean patterns, which can lead, in turn, to additional effects upon the terrain.

For example, during warmer, interglacial eras of terrestrial history, northern continental regions stand higher than they do during colder, glacial periods. This is because in the latter case, they are covered with heavy sheets of ice that weigh them down, while in the former they are lighter and can rise up. However, higher altitudes mean greater exposure to chilly air and strong winds—factors that can lead to snowy conditions, and potentially the formation of glaciers. Once these ice sheets form, they weigh down the land, lowering it, and eventually making it warmer (since lower altitudes generally have high temperatures). In this manner, cycles of ice production and melting might repeatedly succeed each other in a climatic game of leap frog.

Moreover, when the Earth is warmer, it harbors more abundant vegetation, spread out over wider areas. This lush greenery absorbs considerable amounts of excess carbon dioxide, leaving relatively little in the air compared to less bountiful times. As the fraction of atmospheric carbon dioxide diminishes, more of Earth's heat can escape, causing it to cool down. Eventually, as the weather chills, snow builds up, glaciers form, and large, once-forested areas become inundated with ice. Fewer trees mean greater carbon dioxide, leading to increased trapping of surface heat, and ultimately to the end of the frozen era.

Thus, it would potentially take far less than a change in Earth's tilt or orbit to trigger off an ice age. Climatic feedback mechanisms— land movements and alterations in atmospheric composition, for example—might render in mere centuries what astronomical processes would take millennia to carry out. Rapidly ensuing "mini ice ages," such as the Younger Dryas, serve as testimony to the fact that cold and warm spells flash in and out of geological history much more easily than once believed possible.

Changes in Earth's climate that fall far short of ice age conditions but are sharp enough to be noticed can take place in as little time as a year. A strong enough volcanic explosion, for instance, might release enough ash into the air to blanket our planet, block the Sun, and lower global temperatures for an extended period. Such was the case in the early 19th century when the eruption of an Indonesian volcano helped produce what was called the "year without a summer."

Year without a Summer

In 1815, Sumbawa, an island of the Indonesian archipelago, experienced the most powerful volcanic eruption since the end of the last ice age (as far as geologists can tell). More than eighty-eight thousand people perished, either directly or indirectly, when Mount Tambora exploded and spewed its deadly molten lava and ash over Sumbawa and neighboring regions. Though the deadly impact of the eruption was felt most keenly and immediately by Indonesia, its aftermath was experienced the following year throughout the Northern Hemisphere. During that famous "year without a summer,"

Tambora's ash, scattered through the atmosphere, blocked sunlight enough to lower average global temperatures.[3]

Mount Tambora hid its inner violence well. Like a mild-mannered individual suddenly transformed by rage, it burst forth violently and unexpectedly. Thinking that it had long been extinct (or even nonvolcanic), geologists were astonished by its unprecedented potency.

On April 10, 1815, residents of Sumbawa and neighboring islands began to hear loud "cannon bursts": thundering signs of imminent volcanic eruption. A rain of ash began to pour down, lightly covering all regions lying within a few hundred miles of Mount Tambora. By evening, those living in the communities of Sanggar (twenty miles to the east of the mountain) and Tambora (fifteen miles to the south) were mortified to see three columns of flame shooting up into the sky. An ocean of liquid fire, pouring out from what used to be the mountain, presented a macabre spectacle in the nighttime sky.

Consumed by torrential lava or pelleted with pumice stones, thousands of people met gruesome, untimely deaths within hours after the explosion. The eruption literally wiped out centuries of settlement. Subsumed in molten material, the village of Tambora became a mere memory. Sanggar turned into a whirlwind of soaring trees and other flying objects—uprooted by hot, high-speed air currents. By the time the volcano caved in under its own weight, its surroundings had become a veritable graveyard, marked by tombstones of cooling magma.

Even after the eruption ceased, a ghostly cloud of ash hung over Sumbawa. As winds spread the soot through the air, more and more of the region became engulfed in smoke. For many days, the shroud of dust darkened the skies for hundreds of miles around. So many crops were ruined that tens of thousands of people starved to death. These ghastly consequences made Tambora the deadliest, as well as the most energetic, volcanic eruption in history.

The mammoth amount of smoke generated by the explosion gradually dissipated around the globe. By the summer of 1815, unusually brilliant and colorful sunsets were observed as far away as London, a result of the thin layer of ash that filtered the sunlight as it passed through the atmosphere. A portion of the Sun's energy was blocked by the haze, resulting in short term global cooling. Bucking a warming trend that had taken place during the previous decade, average world temperatures dropped about one degree Fahrenheit over the following year.

A single degree of temperature decrease may not seem like much. It was enough, however, to trigger an oddity: a year without summer for parts of North America and Europe.

In 1816, those living in the northeastern part of the United States found themselves shoveling snow in June. As strange as it might seem, more than a foot of powder blanketed that region during that extraordinary month. Summer fashions included scarves and mittens; swimsuits stayed packed away in dresser drawers. Hot weather never came. A freezing autumn succeeded the frigid summer. Autumn was followed, in turn, by a bitter cold winter. Crops failed, and famine ensued in many places. Those who lived through those times experienced, no doubt, a brief taste of what the dawn of an ice age would be like. If the Younger Dryas was a "mini ice age," one can say that they endured a "micro ice age."

Might a year without a summer happen again? Yes, if enough dust filled the skies and blocked out the Sun. Another massive volcanic eruption—or series of eruptions—of Tambora's strength or greater, would probably prove sufficient. The less powerful, but still monstrous, explosion of the Indonesian island of Krakatoa in 1883 generated considerable quantities of airborne soot—enough to be seen around the world, but insufficient to lower summer temperatures.

A summer-free year might have other causes. As discussed earlier, the impact of a large comet or asteroid would also kick up enough dust to blot out the Sun's warmth and lower global temperatures for months or years. The out-of-control fires generated by full-blown nuclear war would similarly fill the skies with smoke, blocking sunlight and cooling down the surface. Plunging the Earth into a frigid twilight known as nuclear winter, planetary ecology would suffer immeasurably and perhaps never fully recover. In the worst case, all life-sustaining resources would be destroyed and the human race (those who survived, that is) would starve to death.

These are all frightening possibilities. If geological history repeats itself, a new ice age is an inevitability. Even if volcanoes, extraterrestrial collisions, and nuclear war do not cool off the Earth, more mundane factors, such as periodic oscillations in weather patterns, and the effects of Earth's position in space relative to the Sun, will eventually bring about another frozen era.

When the next full-fledged glacial age transpires, summer-free years will become the norm—not oddities—in many northern lands. The Tambora experience will be magnified more than a thousandfold. Winter's icy fingers will grasp hold and not let go for many millennia. Under the bone-chilling skies, present-day life will seem a long-gone paradise.

Return of the Glaciers

Someday, perhaps several thousand years from now, snow will once again blanket Canada, New England, Siberia, Scandinavia, the British Isles, and other regions of the north year round. Gradually, thick ice sheets will press upon the soft torso of these areas, and then slowly squeeze the life out of it. Sliding south, these frozen layers will cover more and more of the body of North America and Europe until the day comes when the climate warms and they release their melted waters. New cracks and fissures will open, new lakes will be created, and the Earth's topography will be significantly altered.

Will the human race survive the next ice age? Almost certainly, it will continue, albeit under much less comfortable circumstances. We are a hardy race, used to extremes of temperature and other living conditions. Our ancestors weathered the most recent ice age with far fewer resources than we have today. Therefore, there is no reason to believe that a new glacial era would spell doom for humankind.

Unlike sudden global-cooling catastrophes such as nuclear winter or the impact of an asteroid, the advent of a new ice age would take place quite gradually. Its tortoise-like arrival would allow humankind, along with terrestrial ecology, considerable time to adjust. Chances are, we would be able to survive in some fashion.

Whether or not our civilization could maintain its current living standards under icy circumstances is a different story. Most likely, the advancing of glaciers from the north would force humankind to live within a more restricted region of the Earth—a narrow band hugging the equator, and little more. Agriculture and industry would, in all probability, be forced to operate within this limited zone. Consequently, overcrowding and overtaxing of resources would pose dilemmas of gargantuan proportions, and widespread famine might even ensue. Though the coming of an ice age would not portend armageddon, only a fool would celebrate its arrival.

Imagine a world much colder than today, one in which much of the Northern Hemisphere is covered by ice sheets. Most of Canada is frozen, save a few coastal locales, such as Vancouver and Halifax. Much of the northern part of the United States, from New England, upper New York State, and Michigan to North Dakota, Montana, and Idaho, appears to be heading the same way. Once great cities, such as Minneapolis, Montreal, and Toronto are shells of their former selves. Only by elaborate, costly systems of heated streets and sidewalks, as well as an extensive network of underground passages that connect most of the buildings, have a skeleton crew of citizens there managed to live in reasonable comfort. The inhabitants of those places are lucky glaciers have not demolished their dwellings—at least as of yet.

Farther north, many towns have been abandoned to the powerful ice flows. Alaska, even most of the southern part, is almost completely devoid of people. Few want to live in an icy wasteland, where basic services are absent. The Inuit peoples and other native Canadians and Alaskans have largely become refugees. Though they have been used to cold and isolation, they have reached their limits and escaped to warmer climes.

Wheat and corn are no longer grown in the American prairie. Most of the part south of the glaciers now looks like tundra, where only short grasses and weeds take root. Farmers have sought to relocate farther south. That isn't easy; the price of land in the south has skyrocketed. Moreover, if they try to sell their former farmlands, now barren, there are no takers. No reasonable person would shell out money for wasteland. Consequently, large tracts of property have simply been abandoned.

Hundreds of thousands of Americans and Canadians have streamed into Mexico and Central America in a mass emigration southward. Ironically, Mexican border guards now conduct the daunting task that United States officials used to perform: turning back illegal aliens. In spite of the best efforts of its government, Mexico's own poor citizens have become displaced over time by comparatively wealthy immigrants buying up their land. Rumors of revolution are muttered by a frustrated people increasingly under siege by waves and waves of unwanted guests.

Everywhere—not just in America, but also in Europe and Asia where the cold has hit just as hard—prices of food, clothing, and other essentials have become outrageous. With fewer arable lands, growers must charge more to recoup their costs. The average person would give anything for a fresh loaf of bread or even a palatable piece of crust.

More limited grazing areas mean fewer sheep and therefore fewer wool products, such as sweaters. With warm clothes in short supply—yet greatly demanded because of the frosty weather—they become precious commodities. To those dying in the bitter cold, scraps of fur, pounds of coal, and gallons of kerosene become far more valuable than gold.

In many places around the globe, particularly in the north, life has become solitary, poor, nasty, brutish, and short, as the philosopher Hobbes once said. But it is still life—and must still be lived to the fullest extent possible. Though reality is frigid and bleak, greener days undoubtedly lie ahead. Such is the world's hope in an insufferable new ice age.

Nature's Demons

Nature's power is awesome indeed, cruel and senseless at times, and often unpredictable. It is no wonder that many live in terror of its force, often trying in vain to anticipate the next form its devastation will take. Faced with the random horror of incomprehensibly brutal disasters, such as volcanic eruptions, plague outbreaks, cometary collisions, and floodwater inundations, many frightened individuals and movements turn to the calculus of religious prophesy to try to make mathematical sense out of a chaotic world.

Rather than panic in the face of natural threats, real or perceived, one must work vigilantly to discover workable solutions to these problems. Given enough time and energy, the bulk of potential disasters might be averted through careful planning. Time spent aimlessly worrying about what awful things someday could occur, might better be spent developing the skills to cope with possible catastrophes. Accurate scientific information disseminated to the public at the right time often provides the antidote to despair and the incentive to action.

For example, upon the first outbreak of AIDS in the 1980s, a few doctrinaire religious groups preached that it was a curse—an act of retribution against sinners. Some individuals avoided all contact with anyone they suspected of having the disease, or even shunned cities where it was prevalent. It took considerable effort by scientists and humanitarians (such as the late Princess Diana of Wales, for example) to channel public energy into an understanding of the genuine issues and challenges posed by the illness, realistic methods of minimizing chances of exposure to the virus, ways of truly helping the afflicted, etc. Unfortunately, there is much more to be done as the disease is still spreading in many parts of the world. Educational programs have helped with prevention, treatment, and especially with the reduction of misunderstandings.

The same measured approach would be appropriate for any potential catastrophe. If a threat looms, one must examine how imminent it is, how much damage it might cause, and what might be done to remedy its ill effects. Some situations, such as epidemics, would require immediate attention; others, such as environmental pollution, painstaking efforts over time. Yet other problems, such as new ice ages, would be so slow to transpire they would best be addressed through long-term attempts to ameliorate their effects.

As menacing as natural crises might seem, when monumental forces are suddenly unlocked by human design, it is even more terrifying. Nothing is more frightening than the possibility of nuclear war. In an instant, an atomic explosion could obliterate any one of the world's major cities. Furthermore, if the world's arsenal were fully employed, a shroud of dust would envelop the Earth for months, maybe years. With extraordinarily catastrophic consequences a given, the launching of such weaponry must be prevented at all costs. Over half a century ago, science released the nuclear genie and, unless that bottle can somehow be recorked, humankind will never breath easy again.

CHAPTER FIVE

THE FIRES OF HIROSHIMA

NUCLEAR ARMAGEDDON

For the Angel of Death spread his wings on the blast,
And breathed in the face of the foe as he passed;
And the eyes of the sleepers waxed deadly and chill,
And their hearts but once heaved, and forever grew still!

—LORD BYRON, *"The Destruction of Sennacherib"*

Energy from Matter

I t is one of history's supreme ironies that the equation most commonly associated with the development of atomic weaponry was formulated by the lifelong pacifist Albert Einstein. This association is certainly not justified. Einstein did not anticipate the bomb, nor did his theories aid directly in its development. Yet when many contemplate atomic blasts, they think of Einstein and his famous equation.

Perhaps, if somebody deserves this distinction, it should be physicist Ernest Rutherford, rather than Einstein. In 1904, shortly before Einstein proved mass is equivalent to energy, Rutherford was the first to predict that if atoms could be broken apart great power would be released.[1] Or maybe James Chadwick, who in 1932 discovered the neutron, should also share credit (or blame, as the case many be). Without his findings, atomic nuclei would be little understood. Fault Enrico Fermi, who unknowingly first split the atom in 1934, or Otto Hahn and Fritz Strassman, codiscoverers of nuclear fission in 1938, if you must. Better yet, blame Hitler, whose quest for world power frightened the West into developing the bomb. But don't condemn Einstein.

In his long and fascinating life, Einstein played absolutely no role in the advocacy, design, and implementation of the atomic bomb, except to send a letter to President Roosevelt warning that the Nazis might develop it first. (Even the scientists who developed it called it Hitler's bomb.) Einstein was horrified to see such terrible weapons dropped on Japan. Afterward, it troubled him to ponder that the public associated his discoveries with such horrendous destructive power.

Einstein, who was born in Ulm, Germany, in 1879, spent the waning years of his life (the late 1940s and early 1950s) trying to seal shut the Pandora's box nuclear theory had opened. Yet, once knowledge is gained, it cannot be "unacquired." The human race must forever cope with the possibility that the release of nuclear forces will lead to its annihilation.

To understand Einstein's notion that matter can be freely converted into energy (and vice versa)—the idea that demonstrates how power can be squeezed out of an atomic nucleus—we must grapple with his famous theory of special relativity. It was through painstaking logic and rigorous mathematical deduction Einstein reached the inescapable conclusion that mass and energy form dual faces of the same coin.

Before the age of Einstein, Newtonian physics, with its concept that mass was indestructible, and energy could not be created nor destroyed, held sway for many centuries. Newton's notion that space and time were absolute concepts, independent of the positions and motions of observers, formed the basis of a clockwork physics considered sacrosanct. Einstein's relativity, formulated in the first decade of the twentieth century, demonstrated that, on the contrary, if one makes the experimentally proven assumption that the speed of light is constant, one must then accept mass can be changed, and space and time be altered, when one observer travels faster or slower relative to another.

Let us consider the steps that led to Einstein's deduction. In the late nineteenth century, physicists Albert Michelson and Edward Morley conducted a well-crafted experiment designed to see if light waves, the carriers of energy, appear to move at different rates if viewed at varied observational speed. Their results were negative—one of the most famous null conclusions in science—showing that the speed of light is constant no matter what speed an observer is moving.

Imagine running beside an automobile that is traveling quite quickly. If you were a good enough runner (with "bionic" legs, let's say), and managed somehow to keep pace with the vehicle, it would seem to you to be standing still. Though both you and the car would continue to clock high velocities with respect to stable objects such as lampposts, your speed relative to each other would in effect be zero.

Now let's replace the car with a light signal. The Michelson–Morley experiment proved that no matter how fast you went, if you were jogging beside a light wave, the wave would always seem to be whizzing by at the same speed. Unlike the situation with the car, even if you could run nearly as fast as the speed of light itself, the light pulse would never seem, from your perspective, even an iota slower.

Einstein found this strange. He resolved this apparent dilemma by postulating that if you move fast enough—nearly the speed of light that is—your personal clock slows down with respect to fixed (nonmoving) observers. That's why both a static individual and one traveling at an ultra-high velocity would each record light's speed to be the same. The fast person's watch would slow down as he threatened to catch up to the light's pace. Though, as he hastened his speed, the light would seem to be moving shorter distances, because his watch's second hand would creep along, the light would seem to him to take less time to travel these briefer intervals. Therefore he would continue to perceive the signal as streaking past at an identical rate.

Upon completing his formulation of the principles of relative time and space, Einstein applied them to physical theory, gauging whether or not they led to contradictions. In one such application, he calculated the energy of a high speed particle and then compared his result to that of the same object when still. He found, in contrast to Newton, that even when a body is at rest, it still harbors an intrinsic energy, which he called its "rest energy." He determined an object's rest energy to be proportional to its mass and established the proportionality factor to be the speed of light squared. Like ice can be melted into water, according to this famous equation, mass can freely be transformed into energy.

Those savvy enough to understand Einstein's equation realized it meant the unlocking of great sources of power, that could potentially be used for horrendously effective weaponry or seemingly boundless civilian energy. Even though many believed the atom could be harnessed in principle, few thought that it could in practice.

As late as the mid-1930s, the notion of an atomic blast remained a figment of science fiction writers' vivid imaginations. At that time, even Einstein doubted the atom could be smashed. Reportedly, in 1935, very shortly before artificial nuclear fission was first developed in Germany, he told a press conference in Pittsburgh that the chance of converting matter into energy "is something like shooting birds in the dark in a country where there are only a few birds."[2]

Pandora's Blast

If armageddon had a face it would be Hitler's. No one in history has exceeded in savagery his meticulously orchestrated attempt to conquer Europe and wipe out its population of Jews and others he considered undesirable. The Holocaust provides the quintessential example of how the dark side of human nature, if allowed to flourish unchecked, can erupt into a monstrosity of incomprehensible proportions. The millions and millions of people murdered by Hitler's forces in the 1930s and 1940s (including Russian, American, British, Chinese, and countless other soldiers who died in the process of liberation) would have made the world that much richer if their lives had continued. Instead of celebrating their achievements, we have been forced by tragic history to mourn their lives cut short. Recalling scientists, composers, politicians, children along with their mothers and fathers—so many who are no longer here—the human mind still cannot fathom the horrendous dimensions of such abominable destruction.

Those who followed Hitler's career and witnessed his evil doings during his reign of terror were absolutely convinced that if he came upon a weapon of mass destruction, he would not hesitate to use it. When Austrian scientist Lise Meitner fled the Gestapo secret police in 1938 and arrived in neutral Sweden, she brought word that nuclear fission (splitting apart of atomic nuclei) was being developed back in Germany by her former colleagues, Otto Hahn and Fritz Strassman. Shortly thereafter, she received news from Hahn and Strassman that they had split the uranium nucleus into two roughly equal parts, releasing energy in the process. In January 1939, renowned Danish physicist Niels Bohr, who had been notified by Meitner of the German discovery, startled the scientific community by announcing that fission had been successfully achieved. The weapons potential of this powerful process was obvious to all concerned. Soon, the Second World War broke out, and those in the know became convinced that Hitler would try and might very well succeed in building a bomb based on uranium fission.

Hungarian physicist Leo Szilard, another refugee of fascism, was especially concerned. Along with his colleague Walter Zinn of Columbia University, he had discovered that it was possible to produce a sustained nuclear chain reaction. They had shown that by bombarding an uranium atom (technically a less common type of uranium known as uranium-235) with neutrons, its nucleus would split, generating energy and more neutrons in the process. If other uranium nuclei were near enough, some of the newly created neutrons would split those open as well. Additional energy, and even more neutrons would come forth and could potentially cause further fission. Again and again, this could happen, until the uranium sample was exhausted. For a large enough sample, the process would be self-sustaining (it would keep on going without intervention) producing enormous quantities of power, which might have civilian as well as military uses. Szilard feared the latter, especially in the hands of a tyrant such as Hitler.

In July 1939, shortly after Hitler's annexation of Czechoslovakia, but several months before he began World War II by entering Poland, Szilard contacted Einstein about his concerns. (Physicists Edward Teller and Eugene Wigner also spoke with Einstein about this matter.) He convinced Einstein that Hitler's acquisition of uranium and attempts to develop the bomb must be defeated at all costs. Together with Szilard, Einstein drafted a letter and sent it to President Franklin D. Roosevelt, warning him of the dangers of the Nazis building a nuclear bomb.

Roosevelt responded to Einstein's letter by creating an Advisory Committee on Uranium. The committee issued a report, which urged a "thorough investigation of nuclear physics and the coordination of the work being done at the various universities," which Roosevelt filed "for reference."[3]

Einstein, concerned that no action had been taken, wrote Roosevelt a second letter. The President decided to enlarge his Advisory Committee, and push forward the investigation of atomic weaponry. On December 9, 1941, two days after the Japanese attacked Pearl Harbor and the United States entered the war, he authorized the development of the top-secret Manhattan Project.

Cities of Ghosts

Starting in 1942, Einstein started to notice his best friends in the physics community were disappearing. With a horrific war raging, they obviously weren't going on vacation abroad. Rather most of them had joined a classified project—whose purpose Einstein could easily guess.

The Manhattan Project, headed by American physicist J. Robert Oppenheimer, was one of the largest scientific endeavors in human history. Constituting thirty-seven installations in nineteen states, it employed about forty-three thousand people, and ran on a budget of $2.2 billion—an astronomical sum for those times. New cities were built or expanded, from Oak Ridge in Tennessee to Los Alamos in New Mexico, where Oppenheimer's design work centered. Amazingly, all this construction took place in total secrecy. Not until bombs were dropped, did the public know they had been developed and built.

In July 1945, the first experimental uranium bomb explosion took place in Alamogordo, New Mexico. By that time, the war was virtually over. Hitler had committed suicide, and Germany was occupied by the Allies. Only Japan fought on.

Meanwhile, Roosevelt had died, and Harry Truman had assumed the presidency. Truman wanted to end the war as quickly as possible. He sought advice as to how to pressure Japan for surrender. Some scientists, such as Szilard, urged Truman to warn Japan about the bomb, but not to use it. Others suggested a demonstration of its devastating potential be performed in an uninhabited area, to scare the Japanese into giving up the war. But what if the demonstration failed and the United States appeared foolish?

In the end, Truman decided to drop the bomb over two civilian targets in Japan: Hiroshima and Nagasaki. On August 6 and 9, 1945, the first and last (one hopes) use of nuclear weapons in wartime took place. The bombs met their targets as planned, bringing instant devastation to the population centers. With two of their cities in ruins, the Japanese quickly surrendered.

It is impossible to fathom the magnitude of the devastation as Hiroshima and Nagasaki were each hit by multi-kiloton explosions (one kiloton possesses one thousand times the explosive power equivalent of TNT). The people in the immediate areas where the bombs hit were instantly vaporized. Shadows etched into pavements were their only remains.

Buildings turned into rubble in a flash. Dust, including disintegrated human beings, billowed up into the sky, forming eerie mushroom clouds. Firestorms spread outward from the impact zones, engulfing many square miles of each city.

Survivors of Hiroshima and Nagasaki recall scenes of unimaginable horror. Mothers, holding their babies, were transformed into statues of charcoal. Limbs were blown off and swept through the air. Scalding air currents burned men's faces away, leaving only skulls. Those merely looking at the flash had their eyes instantly melted. Carcasses piled up in streets strewn with broken glass and debris. In short, the terrors of the inferno were unleashed for a time on earth.

Those who lived miles from the devastated city centers were still in grave danger. Winds, of many hundreds of miles per hour, carrying radiation from each blast, brought slow painful deaths to hundreds of thousands of victims. Radiation poisoning, leukemia, and other forms of illness plagued the bombed regions for years.

The majority of those who physically survived suffered from deep psychological ailments, ranging from misplaced guilt to psychosis. Many questioned why they had outlived their loved ones.

Others became obsessed with cancer, fearing that the consequences of radiation exposure could appear at any time. Recurrent nightmares, panic attacks, and other symptoms of stress were ubiquitous.

Upon learning of what happened to the Japanese, many of those involved with the bomb's design felt sincere regret. Oppenheimer, who had supported the bombings, was greatly affected by their horrific consequences, and became stricken with remorse. Many scientists urged an end to further development of nuclear weapons. By then, however, the fate of the bomb was beyond their control; politicians, such as Truman, governed its destiny.

Naively, experts in the American military thought it would have a monopoly on nuclear weapons for decades. They were proven wrong shortly thereafter when the Soviets tested their own atomic explosives. Soon, the United States and the Soviet Union were locked in an expensive arms race—the Cold War—that drained both economies for decades.

As the years rolled by, the weapons of mass destruction grew more and more powerful. Thermonuclear (hydrogen fusion) bombs, introduced in the 1950s by Andrei Sakharov and his colleagues in the Soviet Union, as well as by Edward Teller and others in the United States, generated explosions in the megatons. Thus they had thousands of times the destructive capacity of the uranium-based kiloton bombs.

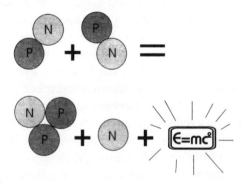

The process of fusion. Depicted are two deuterium (heavy hydrogen) nuclei, each consisting of a proton and a neutron, releasing a neutron as they combine to form helium-3. Following Einstein's prinicple of mass–energy equivalence, matter is converted into energy in the process.

In a thermonuclear (hydrogen fusion) weapon, a fission bomb, containing uranium and/or the artificial radioactive substance plutonium, is used as an explosive trigger to squeeze together hydrogen nuclei and form helium. Each helium nucleus (two protons and two neutrons) has twice the number of particles as a heavy hydrogen nucleus (one proton and one neutron), but less than twice the amount of mass. In accordance with Einstein's principle, the mass excess during fusion becomes transformed into radiation. As myriads of hydrogen atoms are fused, enormous quantities of energy are released, creating a much greater explosion than the initial fission blast that triggered it.

A single thermonuclear device, if exploded over a metropolis, would turn that city instantly into ruin and render the surrounding countryside lifeless. Respected physician Helen Caldicott has detailed the devastating impact of a twenty megaton bomb on a major city:

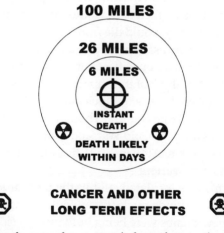

If a twenty-megaton thermonuclear weapon is dropped on a major city, some scientists estimate that all those living within six miles of the blast would be annihilated immediately. Those living farther away, but still within tens of miles from the epicenter, would be subject to severe burns, radiation poisoning, the impact of collapsing buildings, and other life-threatening effects. Most within this zone would die within days. Even farther away from ground zero, millions would likely develop cancer and other long-term medical hazards.

If it explodes at ground level on a clear day it will release heat equivalent to that of the Sun, several million degrees Celsius, in a fraction of a millionth of a second . . .

Six miles from the epicenter, every building will be flattened and every person killed. . . . People just beyond the six-mile, 100-percent lethal range who happen to glance at the flash could have their eyes melted . . .

Twenty-six miles from the epicenter, the heat from the explosion will still be so intense that dry objects such as clothes, upholstery, and dry wood will spontaneously ignite. People could become walking, flaming torches. . . . [4]

Even after the heat from the blast has died down, millions would be exposed to massive doses of radiation, at levels well above the lethal quantity. For those bathed in radioactive fallout, death would follow a tortuous path. Excruciating pain, severe nausea, vomiting, and bloody diarrhea would all occur within days—the time of onset depending on the amount of exposure. Victims would become tired and weak; their skin would burn and their hair would fall out. Ultimately, they would die from internal bleeding, infection, dehydration, severe burns, or perhaps from a combination of factors.

Considering the ruinous effects of a single thermonuclear devise, one shudders to think of the impact of many bombs launched at once. Scientists estimate several thousand of them, deployed in unison, could destroy a large country. Though stockpiled in the tens of thousands throughout the Cold War, and tested extensively (the first was detonated on Bikini Atoll in 1952), these new bombs were fortunately never used in warfare.

As the arms race escalated, all but the most ardent war hawks realized that a worldwide nuclear war could not truly be won. The only reasons new bombs were being built by each side were to keep up parity with the other side, and to try, if possible, to obtain superiority. Each party to the conflict knew, however, that a full scale nuclear war would represent the downfall of civilization, east and west. Therefore, neither side desired war; it just wanted the other side to fall behind in the arms race, lose military and political leverage, ruin its economy, and ultimately give up.

Einstein lobbied hard for the halting of the arms race, which he viewed with repulsion and despair. At every opportunity, he pointed out the hazards of continuing to embark on such a course: increased militarization, growing secrecy, and police control, as well as an atmosphere of general hysteria. His solution was a strong, democratically-administered world government that would administer a general disarmament.

On April 11, 1955, only one week before he died in Princeton, a frail Einstein affixed his signature to a public appeal known as the Einstein–Russell Manifesto. Drafted by influential British philosopher and pacifist Bertrand Russell, the petition called for international cooperation between scientists from East and West in an attempt to end the arms race. The appeal was signed by other leading scientists from the West, and resulted in the convening of the Pugwash Conference on Science and World Affairs, attended by delegates from both sides of the conflict. Nevertheless, the scientists' words went unheeded, and the escalation of nuclear weapons continued.

Though the arms race seemed (especially by the 1970s and 1980s) to have become a foolish, and very costly game, rather than a genuine prelude to war, many pondered what would happen if the weapons stockpiles—of tens of thousands of devices—really were used. Might the world really end? While director Stanley Kramer explored the tragic aspects of this possibility in his film, *On the Beach,* filmmaker Stanley Kubrick examined its darkly comic side in *Dr. Strangelove.* In both influential movies, the launching of the world's atomic weapons heralds its cataclysmic demise—an all too real possibility. The term "doomsday device," used in Dr. Strangelove to describe a bomb of ultimate destructive capabilities, became a vivid symbol in the mind of the public of what apocalyptic dangers might lie ahead.

In the 1970s, with environmentalism a popular cause, many scientists pointed out that nuclear war would represent the ultimate environmental nightmare. Vocal opponents to war spoke with dread of the direct effects of the blast, as well as of subsequent damage due to radiation poisoning. However, they forgot one critical factor: smoke. A nuclear war would churn so must dust into the air that, as a group of five reputable American scientists pointed out in 1982, a cold dark winter would last for months, maybe years. Nuclear winter, they called it—one of the gravest threats imaginable.

Cold Silence

The idea of nuclear winter was discovered accidentally. In 1971, American astronomer Carl Sagan, renowned for both his provocative scientific ideas and his superb popularization of such notions in books and on television, was working on resolving problems facing the Mariner 9 mission to Mars. Mariner 9, the first spacecraft sent into orbit around another planet, was designed to last for only three months. During that span, it was supposed to relay photographs and other information detailing conditions on Mars back to Earth for analysis. The dilemma Sagan and other mission observers faced was that the red planet was in the middle of a lengthy dust storm. For many weeks, photos taken of the planet looked just like a blank disk. No features of the Martian surface could be seen through its dusty shroud. Sagan and his colleagues started to wonder if any useful data at all would come out of the mission. Fortunately, the dust cloud did gradually dissipate before the spacecraft lost function.

One of the interesting pieces of information that did get sent back to Earth was the ever-changing Martian surface temperature. When the planet was fully enveloped in dust, Mars was much colder than it was after the dust eventually settled. Sagan, along with colleagues (and former students) James B. Pollack, O. Brian Toon, and Thomas Ackerman, developed a computational model of how atmospheric dust content affects surface temperature.

Then, in 1980, Luiz Alvarez advanced his famous hypothesis that the dinosaurs and other species on Earth were wiped out by the impact of an extraterrestrial object. Dust from the collision filled the air, darkening the skies, cooling the terrestrial surface, and creating a false winter that lasted for many months. Ultimately this led to the extinction of many types of living beings.

One year later, Paul Crutzen of the Max Planck Institute for Chemistry in Mainz, Germany, and John Birks of the University of Colorado performed a study estimating the amount of smoke pro-

Carl Sagan, 1934–1996, Cornell professor of astronomy. (Courtesy of Cornell University)

duced by the firestorms burning forests and cities during a nuclear war. They calculated how must dust would rise into the air and darken the skies.

Sagan, Pollack, Ackerman, and Toon, along with Richard Turco of R&D Associates, a private research company in California, decided to apply their atmospheric dust model, gleaned from the Martian study, to estimating the effects smoke from a nuclear war would have upon world climate. Working at Ames Research Laboratories in Moffett Field, California, the group, collectively nicknamed TTAPS (for Turco, Toon, Ackerman, Pollack, and Sagan), developed startling new visions of the aftermath of global thermonuclear conflict.

At an international meeting in Washington, D.C., "The World After Nuclear War, the Conference on the Long-term Worldwide Biological Consequences of Nuclear War," held October 31, 1984, Sagan presented the TTAPS results to more than five hundred concerned participants. During the opening address, Sagan, pointing out that it was Halloween, detailed his own apocalyptic "ghost story," of the ghastly months after global thermonuclear conflict.

In a World War III scenario, thousands of missiles would be launched, carrying many thousands of warheads. A typical nuclear warhead, aimed at a city and set to explode above it, would create an initial fireball more than ten miles in radius. Within this region, the most flammable materials would ignite first, setting off smaller fires. Soon, a firestorm would envelop the city, burning down houses, churches, businesses, and schools, and spreading more and more. Eventually, the entire region would experience a massive, hellish conflagration, that would pump hundreds of thousands of tons of smoke and soot into the sky.

Imagine such a fiery ruin, happening simultaneously all over the globe, and picture how much burnt material would fill the skies. TTAPS estimated a global nuclear war, involving less than half of the superpowers' weapons at the time, would generate more than 100 million tons of smoke, rising high up into the stratosphere. Consequently, more than 95 percent of the Sun's light would be blocked for months.

As in the case of the Martian dust storms, the TTAPS team predicted the smoky shroud over Earth would darken its skies and dramatically cool down its climate. A wintry false twilight would last for hundreds of days. Under those frigid conditions, with photosynthesis impossible, the bulk of the world's crops would likely wither and die. Lacking basic sustenance, most animals would perish as well.

In short order, millions of people would starve to death. Stomachs empty, the desperate populace would scrounge for food, but virtually none would be available. With the decimation of most edible plants and animals, the human race would possibly face its own extinction.

To make matters worse, a hard rain of radiation would continually beat down upon the planet. Those who survived the blasts and food shortages would be condemned to remain in shelters forever, rather than facing exposure to lethal rays. From radiation poisoning and cancer, to perpetual darkness and cold, life on Earth would be brutal, if not impossible.

The Late Great Planet Earth

As we have seen, whenever calamity (or perceived threat of disaster) occurs—be it through floods, plagues, comets, or other agents—many turn to religious teachings for solace. Some of them become drawn to apocalyptic movements, as we discussed earlier. The prospect of nuclear catastrophe is no exception to this rule.

In 1970, evangelist Hal Lindsey published what has become the most popular doomsday manifesto of all times, *The Late Great Planet Earth*. Selling millions of copies, Lindsey's book epitomizes the apocalyptic way of thinking in the late twentieth century, in much the same manner Miller's writings reflected nineteenth century viewpoints.

Lindsey viewed two historical events as critical to an understanding of biblical eschatology: the development of the nuclear bomb, and the foundation of the State of Israel. He was hardly alone in seeing these as defining moments. When the bombs exploded over Hiroshima and Nagasaki, the world could not help be reminded of the vivid descriptions of armageddon that lie in the last book of the New Testament. Who could dismiss thoughts of "fire and brimstone" from their minds?

Three years after the Second World War passed into history, the modern State of Israel was declared. Not only Jews celebrated their return to their biblical homeland. Many Christians viewed the Jewish homecoming as a divine miracle. Some even saw it as a symbol of the biblically ordained final days of the world.

The Temple of Jerusalem, mentioned in the book of Revelation, was destroyed by the Romans in the first century A.D. It would need to be rebuilt for New Testament prophecies to come true, some Christians argue. Therefore, by reconstructing the temple on its ancient site—now within Israeli soil—the Jews could play a role in establishing the conditions for the Second Coming.

One of the holiest shrines of Islam, the Al-Aqsa mosque with its celebrated Dome of the Rock, stands today upon the Temple mount. Destroying the mosque (an abominable action that virtually no one would advocate) and rebuilding the Jewish Temple on that site would almost certainly lead to interfaith warfare. Some modern Christians predict such an ensuing conflict would turn global—corresponding to the ultimate battle described in Revelation.

Lindsey, who left his position in Campus Crusade for Christ to research and write his book, sought to unify, explain, and expand these themes. He developed an elaborate compendium of what he saw as prophesies fulfilled, and strived to extend these forecasts into the near future. In his view, the Bible is literally correct, and provides a perfect past and future chronology of the human race. Thus by learning its system, he felt that he had unlocked its lexicon of prediction. He presented his interpretations unabashedly to the reader as the way things certainly will be.

In Lindsey's view, the centuries of world history that occurred between the age of Christ and the foundation of Israel represented a long irrelevant pause in the human chronicle. Like a patient on strict bed rest, in terms of prophesies come true, we practically slept through that period.

In 1948, when the Jewish state was founded, the divine alarm clock rang loudly. The generation living through that momentous occasion would witness all of the events in Revelation come to pass, Lindsey argued. God promised the Jewish people they would return to Jerusalem. Once that promissory note was honored, God could fulfill other biblically-predicted tasks.

Lindsey preached, and continues to maintain, that armageddon is imminent, and will be fulfilled through global nuclear warfare. In 1970, when he wrote his bestselling book, he predicted the rebuilding of the Temple in Jerusalem would trigger worldwide warfare, centered in the Middle East, between the Soviet Union and China, against a united Europe and a significantly weakened United States. (He saw the European common market, ten countries at the time, as the ten-horned, seven-headed beast described in the book of Daniel.) Today, a dozen books later with his updated collection of prophesies, *Planet Earth 2000 AD: Will Mankind Survive?*, published in 1996, Lindsey considerably alters his predicted timetable and players, but not his end day forecasts of doom. The spread of AIDS, the threat of water shortages due to overdevelopment and pollution, and the rise of radical Islamic fundamentalism have become Lindsey's new omens that our time on Earth is limited.

It is not hard to understand why Lindsey saw the need to revamp his prophecies. Between the 1970s and 1990s, the world changed considerably. The Soviet Union and the Warsaw Pact (nations allied to the Soviets) completely collapsed. In their place, are a number of boldly independent nations, Russia the largest, none of which seek confrontation with the United States. Wars have occurred, but have been mainly local, and, in a few cases, regional. None have threatened to engulf the world in battle. With this new perspective, the perils of nuclear weapons do remain, but have significantly altered in scope.

Doomsday Clock

The clockface emblem of the *Bulletin of Atomic Scientists* provides a tangible symbol of humankind's ability to destroy itself in nuclear war. At the dawn of the atomic age, when Einstein and other concerned scientists founded the *Bulletin* in 1947, its symbolic Doomsday Clock was set at seven minutes to midnight—how close the world seemed to be to destruction. Whenever political events have brought civilization to the precipice of doom—such as when the U.S. and the Soviets tested their first hydrogen bombs—the clock's minute hand has been moved closer to midnight. The *Bulletin*'s editors determine exactly where the hand is placed, based upon what they see as the chances of global conflict. Sometimes it has been as close as two minutes before the hour of doom. In contrast, positive developments, such as partial testing bans and arms reduction treaties, have justified pushing it backward. The end of Cold War in the 1990s, the break-up of the Soviet Union and the dramatic reduction in tensions between the United States and Russia have permitted the minute hand of the Doomsday Clock to be placed farther back—at least for a time—than ever before. From 1991 to early 1998, the clock's settings have ranged from fourteen to seventeen minutes to midnight (the least menacing since the *Bulletin* was founded), symbolically depicting the world's great sigh of relief that we no longer seem on the brink of war.

Nevertheless, in spite of more relaxed international relations, nuclear catastrophe still represents a grave potential danger. In June 1998, recognizing this fact and acknowledging the ominous testing of devices by India and Pakistan, the Board of Directors of the *Bulletin* moved the minute hand of the Doomsday Clock to nine minutes to midnight. Tens of thousands of thermonuclear weapons still remain in silos and submarines, mainly in the United States and Russia, but also in China and several other countries. Though in 1994, the two main nuclear powers signed an agreement to cease targeting each other's city and facilities with missiles, targets could be chosen and weapons fired at a moment's notice. Of the twenty-seven thousand nuclear warheads that presently exist, the United States and Russia keep a combined total of more than five thousand of them ready to launch within a few minutes upon orders.

Accidental launchings, and even accidental full blown nuclear war could still happen, even in these days of relative peace. A 1995 incident, which security experts recently brought to light, demonstrates how rapidly a false alert might place the world on the brink of war.

In January of that year, a six-ton rocket was launched from the coast of Norway, three hundred miles from the Russian border, on a routine scientific mission: to study the Aurora Borealis (Northern Lights). Paul Kelly, a Cornell University professor who designed some of the equipment, reports he and his colleagues were careful to notify the Russians about the time, purpose, and trajectory of the launch, just in case of mishap.

Unfortunately, a bungle did occur. High winds blew the rocket off course, directly toward Russia. For unknown reasons, the Russians did not know about the launching—someone had forgotten to record it—and mistook the rocket for a Trident missile. Trident missiles harbor up to eight nuclear warheads and represent one of Russia's worst fears. A single Trident, launched at St. Petersburg, could destroy that city.

Naturally, the Russians did not take any chances. Their military forces immediately went on red alert. President Boris Yeltsin was handed the "nuclear suitcase"—electronically enabling him to "push the button" and launch their arsenal. Within thirty minutes, however, they realized the rocket wasn't heading toward them. The general alert was lifted and the crisis was over.

What if a cool head like Yeltsin wasn't in command? Might a brash leader, on either side, someday trigger off a global thermonuclear war on the basis of a stupid misunderstanding—such as a scientific mission off course, mistaken for a threatening missile? Until all nuclear weapons are dismantled, or prevented from being launched so quickly, civilization must live under a veil of uncertainty.

Though the world's nuclear weapons are kept under tight security, terrorists represent another potential threat. It is frightening to imagine what would happen if a violent, confrontational group decided to express their venom by launching an acquired or constructed missile. To decrease such dangers, stockpiles of nuclear weapons and bomb-making materials, such as plutonium, must be reduced and eventually eliminated (encapsulated into impregnable, well-monitored burial sites, perhaps).

Though nuclear war still remains a realistic possibility, no one would deny that its threat has been reduced. Therefore, many of those concerned about the state of our planet have redirected their activism into new environmental causes. Chief among those are the issues of global warming and ozone depletion—two hazards drastically affecting the Earth's atmosphere and climate. Unchecked, these problems could lead to unprecedented catastrophe: sweltering, crop-destroying temperatures, raging floods, and increased exposure to lethal radiation—perhaps even the destruction of civilization.

CHAPTER SIX

THE SKIES OVER ANTARCTICA

ENVIRONMENTAL NIGHTMARES

What is happening . . . is part of a global pattern. Glaciers are retreating worldwide. This is strong evidence of global warming over the past century—the disruption of our climate because of greenhouse gas emissions into the atmosphere, all over the world. The overwhelming evidence shows that global warming is no longer a theory—it is a reality.

—Vice President Al Gore (*Address at Glacier National Park, September 1997*)

Smoggy Skies

It is said that in Victorian times, the fog was so thick in London it resembled pea soup. British explorers, returning from expeditions through thick rainforests, would find themselves equally challenged by casual strolls through Hyde Park. Big Ben's countenance, often masked by sooty air, could hardly be relied upon as a faithful landmark. London was a murky labyrinth—"the big smoke," it was called, which, on bad days, had very little for sightseers to see.

It does not take Sherlock Holmes to solve the mystery of why London was so smoggy. Coal was the culprit. Coal is an abundant natural resource and a cheap source of fuel. During the 1970s oil crisis, coal was even touted as the answer to the world's energy needs. Yet when it is burned without proper filtration, it gives off a thick, acrid smoke, darkening the skies and filling the air with a sickening aroma.

London was not the only heavily polluted city throughout much of the nineteenth and twentieth centuries. Industrial communities around the world, such as Pittsburgh, Detroit, Cleveland, Dortmund, Glasgow, and Manchester, to name a few, became notorious for their hazy skies, dirty streets, and foul waters. Only in the past few decades have clear, clean vistas returned to many of these places. Pittsburgh, for example, is now known for its magnificent tall buildings, framed against blue skies, rather than for its smog.

Indeed, in much of the developed world, steps taken in the 1970s and 1980s to clean up the air and water have had notable effect. Many pundits who once forecast imminent environmental doom can now take pride their warnings have had strong impact. For the first time in generations, fish and other wildlife have returned to many urban-abutting rivers and lakes.

In some ways environmental regulation is a grand success story. In other ways it is not. While many bodies of water that border highly populated regions, from Lake Erie to the Thames, are cleaner than they were decades ago, the growth of cities and the spread of suburbs have created novel sources of contamination. New forms of waste, from plastic containers to disposable diapers, have overwhelmed disposal systems. Recycling has become more common, but it cannot eliminate the problem, only reduce it. Unless society scales back (and who would willingly give up their luxuries), trash heaps will likely continue to overflow.

Growth of population has meant expansion of population centers. Forests, wetlands, and other wildlife habitats around the world are vanishing, bulldozed over by industry, commerce, housing complexes, and large-scale farming. For every natural area under protection from encroachment, there are countless more silently surrendering to the forces of development. Though its tentacles have been beaten back in some areas, the multilimbed monster of pollution is still vital, constantly eyeing new victims to envelop.

The air seems cleaner in London and New York, but what about Mexico City, Shanghai, and Calcutta? As stricter environmental regulations have forced factories to clean up their acts in the wealthier nations, economic pressures have precluded (or at least slowed down) such reform in many of the less-developed parts of the world. In many places where relatively dirty sources of power, such as unprocessed coal and kerosene, are abundantly used, switching to cleaner fuels, or at least using filters, is considered too expensive a transformation to make. Unfortunately, the costs of maintaining inefficient power systems are usually not weighed against the more subtle expenses of hospitalization of segments of the population (es-

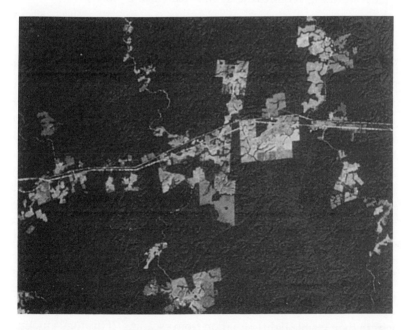

This photo is an image of the Amazon rainforest in Brazil taken by the Landsat-5 satellite in 1990. The light gray regions represent areas recently deforested. Areas of undisturbed forest—less than 85 percent of the Amazon Basin—appear as dark gray. (Courtesy of NASA.)

pecially elderly) for respiratory problems. Some strides have been made in recent decades, but atmospheric pollution is still a problem of critical proportions in much of the globe.

Moreover, environmental scientists have come to realize some of the substances released into the air by human activities though innocuous in the short term, have devastating effects in the long term. This has raised some delicate political issues such as convincing nations—especially in the West—of the dangers of the constant global release of "pollutants" considered harmless in smaller quantities. It is relatively easy to make the argument, for example, that carbon monoxide is poisonous, and its presence in factory fumes and automobile exhausts should be minimized. However, to reduce atmospheric carbon dioxide, a wholly benign substance exhaled in breathing by all animals, but that in tremendous doses is threatening our planet with global warming, requires much more convincing proof.

Message from Kyoto

In 1997, representatives from more than 160 nations gathered in Kyoto, Japan, to discuss a matter of grave importance to the human race: global warming. These delegates were faced with the challenge of negotiating a plan to reduce the production of gases harmful to Earth's climate. The program would need to be reasonable enough to convince their governments it would not bring economic ruin. As Melinda Kimble, United States State Department official and member of the American delegation reported, "These are among the most complex set of negotiations the nations of the world have ever taken."[1]

For more than a century, science has known carbon dioxide and other atmospheric gases affect global climate. As Swedish chemist Svante Arrhenius pointed out in 1896, atmospheric carbon dioxide acts like the windows of a greenhouse, continually letting in sunlight, but blocking heat in the form of infrared (slightly higher wavelength than visible light) radiation from escaping. If the total amount of carbon dioxide goes down, then greater amounts of heat escape from our planet's surface, and global temperatures drop, eventually resulting in an ice age. If, on the other hand, the atmospheric carbon dioxide content increases, then more surface heat becomes trapped, and temperatures increase worldwide. The latter is called the greenhouse effect.

Earth requires a certain balance of carbon dioxide in its air to maintain the temperate conditions favorable to life. If our atmosphere suddenly lost all of its carbon dioxide content (a highly unlikely scenario), then global temperatures would drop so drastically many forms of life would become extinct. (Not to mention most plants require carbon dioxide to grow, but that's a different story.) However, if Earth's atmospheric balance were upset by the infusion of drastically increased quantities of carbon dioxide and other so-called "greenhouse gases," such as nitrous oxide, methane, and ground-level ozone, then it would ultimately become an equally inhospitable hothouse. The results of such a greenhouse effect can be seen on our neighboring planet, Venus.

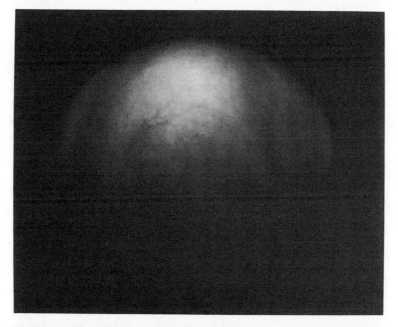

The planet Venus, as taken by the Galileo spacecraft in 1990. The scalding conditions on its planetary surface demonstrate the impact of a runaway greenhouse effect. (Courtesy of NASA.)

Venus' size and mass are so similar to Earth's it has been called our "sister world." Yet the sweltering conditions on our neighbor are so different from our own moderate weather that mistaken identity would be impossible. Though Hades would feel quite comfortable on the second planet from the Sun, nobody else would.

Astronomers believe Venus was once very much akin to Earth in climate, as well as size. Billions of years ago, perhaps, its surface may have been streaked with flowing rivers emptying their watery cargo into magnificent wide oceans. While situated close to the Sun, its average temperatures have probably always been hotter than ours; in the past, its weather may have been warm, but still liveable.

Today, Venus is a hellish desert where no sane creature dare tread. Conditions there are about as hostile as they come. Its surface temperatures range from a sweltering 700 degrees Fahrenheit on its mountain tops to an impossible 900 degrees in its valleys. All surface water has long ago evaporated, leaving the soil bone dry. The current chances of life on that world are comparable to those of finding all the gold of Fort Knox stacked on one's doorstep as a Christmas present. Even the casino industry, comfortable with Nevada summers, has refused to invest there.

Sometime in Venus' past, carbon dioxide, sulfur dioxide, and other heat-trapping gases began to build up in its atmosphere. Over time, they acted to pump up surface temperatures considerably, until virtually all of the water was expelled. Deprived of that fluid's cooling properties, temperatures rose even more, until the present-day inferno came into being.

Might Earth follow the path of its planetary neighbor? Though it is highly unlikely our world will ever resemble Venus, if enough carbon dioxide and other greenhouse gases choked our atmosphere, conditions could become sweltering.

Scientists once believed Earth's atmospheric content could scarcely be affected by human activity. When Arrhenius found accumulation of greenhouse gases could lead to global warming, he little suspected the process was already at work. This attitude changed in the 1950s when data collected by researchers around the world reflected unmistakable long-term increases in the atmospheric percentage of carbon dioxide. (The English meteorologist G. S. Callendar detected this trend in 1938, but his warning went largely unheeded for two decades.)

The 1950s were a time of great prosperity in the West. Bouncing back from the hardships of the Second World War, America and Europe were brimming with vitality. The spirit known in Germany as the "economic wonder" reverberated around the world in an era of growth and affluence. Thousands of newly built or redesigned factories, freed of the burden of producing war items (for a "hot" major war, at least) contentedly churned out their consumer products. Little did anyone know that these industrial plants were also bellowing out gases that were slowly raising the mean temperature of the Earth—and that they and their predecessors had been doing so for decades.

By the end of the decade, the world finally awoke to the problem. In 1957, American oceanographer Roger Revelle warned of the ill effects of greenhouse gases from the burning of fossil fuels—coal and oil, in particular. About the same time, measurements taken by Charles David Keeling on top of the Mauna Loa volcano in Hawaii indicated carbon dioxide levels were rising even in areas far from pollution. Ten years later, Syukuro Manabe and Richard Wetherald of the Geophysical Fluid Laboratory in Princeton, New Jersey, calculated average global temperatures would increase by an estimated 2.5 degrees Celsius (4.5 degrees Fahrenheit), if carbon dioxide levels rose to double their preindustrial values (as they are projected to do within decades at the current rate of increase).

An atmosphere a few degrees hotter may not seem like much, but the implications of such a worldwide temperature rise would be extremely serious. Heat stress deaths would increase significantly, as summer temperatures in currently temperate regions would skyrocket. Weather in Washington, D.C., would become as hot as what Houston now experiences, while Houston would seem like the present-day Australian outback. As temperatures rose, smog would become much more common, pollen counts would climb, and warm climate diseases, such as malaria, cholera, and yellow fever, would move further north.

Imagine walking through the streets of Chicago during a sweltering summer of the future. Ignoring the public health warnings, you step outside for a bit of exercise. Pleased to be outdoors—a rarity—

you seem fine at first. Within minutes, however, baked in the blistering sun, you start to feel light headed. A flashing electronic thermometer (and emergency phone) strapped to your wrist warns you the temperature has climbed above 120 degrees Fahrenheit. Swatting at rogue mosquitoes gnawing at your neck. Praying they don't carry deadly illness, you feel increasingly faint. A queasy feeling fills your stomach, thirst pervades your throat, and you start to choke. Gasping for breath, you find it increasingly difficult to inhale the hot, smoggy air. Disgusting. Experiencing the dizziness that comes with heat exhaustion and oxygen deprivation, you are on the verge of collapse.

Fortunately, you have reached the Lake Michigan shore. Diving in to the water (a "cool" 80 degrees in temperature) momentarily revives you, just in the nick of time. Meanwhile, your wrist device has automatically phoned for help. That's the last stroll I'm going to take, you resolve, as an emergency crew pulls you out of the lake and carries you on a stretcher to a waiting ambulance.

During a greenhouse era, not just the air and land would be affected. In fact, the oceans would probably undergo the most noticeable change. Sections of the Arctic and Antarctic ice caps would slowly melt, pouring their meltwater into the sea. Meanwhile, following the physical principle that warmer objects enlarge, the heated oceans would expand. All this would result in a rise in sea level around the globe. Depending on which predictive model one uses, assuming a doubling of atmospheric carbon dioxide levels, estimates of the rise range from six to thirty-seven inches. Even after temperatures stabilize, the oceans would continue to expand due to their stored heat. Ultimately, the sea level rise could even exceed forty inches—or more.

Studies have shown that if all of the ice in Antarctica thawed, then the world's ocean levels would rise by over two hundred feet. No one truly expects that to happen, at least in the near future; the Antarctic ice cap is far too stable to allow for such a drastic melting. But even a few feet increase in sea level would bring on dramatic changes around the globe.

Such an inundation of ocean water would greatly affect low-lying coastal regions, as well as small islands. Regions just above sea level, such as southern Florida, would undoubtedly be hard hit. The Everglades, much of which is less than twelve inches in elevation, would eventually be subsumed into the Gulf of Mexico. Before it vanished, its ecology would be ruined by the intrusion of sea water. Many southeastern cities, such as Miami and New Orleans, would have to fight for survival. If flooding became bad enough, the buildings of the French Quarter would be soaked up to their second floor balconies; Mardi Gras and other time-honored traditions of that once-lively city would be no more. Eventually, low-lying states such as Louisiana, Florida, and Mississippi wouldn't even exist perhaps, their residents turned into refugees.

A number of small island states around the world foresee life or death battles if global warming continues. Many of the nations of Polynesia, the Caribbean, and elsewhere lie barely a foot or two above sea level in parts. A continued rise in ocean levels could easily drown much of their lands. Unlike coastal residents on large continents, who could conceivably, though painfully, move inland, natives of places such as Samoa, Tahiti, the Cook Islands, and Barbados would have no other choice but to abandon their homelands. Otherwise, they would be fighting off floods in vain, until they were consumed by raging ocean waters.

To combat the frightening possibility that the greenhouse effect could lead to their countries' destruction, a number of island nations have banded together in the Alliance of Small Island States (AOSIS). Though they differ greatly in language and heritage, these peoples are united in their drive to prevent further global warming. Formed in 1990 during the Second World Climate Change Conference in Geneva, the alliance consists of thirty-six member countries and several observer nations fighting for commitments from polluters for reductions in greenhouse gas emissions.

One might think the nations of northern Europe, not exactly the most clement of places, would welcome global warming as a way to enjoy milder winters and toastier summers. Residents of Norway, Denmark, Sweden, and Great Britain might imagine themselves sipping iced tea while sunning themselves on the beach in March or April. On the contrary, they should be out buying parkas, not swimsuits.

According to experts from the Geophysical Fluid Laboratory, the partial melting of the Greenland ice sheet, along with other changes due to the greenhouse effect, would send so much fresh water into the North Atlantic the gulf stream could potentially be diverted. The salinity of the ocean north and west of the European coast could decrease, resulting in greater ice formation there. This could push warmer waters much farther south than they are at present. Conceivably, Europe could lose access to the balmy currents now sweeping its shores.

As we've discussed, many meteorologists believe such a shifting of currents due to freshwater influx from melting glaciers caused the Younger Dryas, the last mini ice age in Europe, over ten thousand years ago. If such a change were to recur, weather in Ireland, England, the Netherlands, and other coastal European nations would become much more akin to that of Canada or Minnesota. Snowmobiles, not schooners, would be the new recreational vehicles of choice.

Europeans looking for discounts on secondhand snowmobiles might wish to phone their friends in Alaska. In coming years, many Alaskans might find theirs obsolete. There, global warming is gradually melting the frozen ground called permafrost, never more than several degrees below freezing, in much of the state. Shifts in warm air currents there have added to the greenhouse effect temperature rise, producing dramatic alterations in climate. Negative consequences, due to these changes already, include the collapse of roadways built on top of permafrost, as well as the proliferation of insect pests. These repercussions are expected to magnify as world temperatures climb. Of course, things won't all be bad; many Alaskans will enjoy their new balmy weather. It won't be long before Alaska becomes known as the new surfers' haven; just make sure to bring enough bug spray.

Because of the stark impact of global warming, a number of international meetings have been held to address the issue. These date back to the First World Climate Conference in 1979, sponsored by the World Meteorological Organization of the United Nations. Back then, at least according to public perception, the issue had just surfaced and didn't seem critical.

By the time of the 1990 Second World Climate Change Conference in Geneva, popular awareness of the greenhouse effect had skyrocketed. The reason: the 1980s constituted the warmest decade on record, with seven of the eight warmest years in meteorological history up to 1990. Records indicated 1987 was the warmest year to date (it has since been surpassed by 1995, the hottest yet). The scorching weather of the 1980s, accompanied by droughts in many places, shocked the world into action. In 1992, 154 nations signed the Rio agreement, limiting the release of carbon dioxide and other greenhouse gases. Though emissions targets were set, they had little effect. Industrialized countries, the United States being the biggest culprit, continued to pump massive amounts of temperature-raising pollutants into the atmosphere. Environmental groups, such as Greenpeace, as well as political organizations, such as AOSIS, pushed relentlessly for stricter, binding standards.

The 1997 United Nations-sponsored climate change meeting in Kyoto was seen as a perfect opportunity for the industrial nations to prove they are mending their ways. With President Bill Clinton and Vice President Al Gore in the White House, self-avowed environmentalists (Gore even wrote a best-selling book on the environment, entitled *Earth in the Balance: Ecology and the Human Spirit*), perhaps a treaty could be orchestrated that would stick. Maybe global doom could be averted after all.

Alas, Clinton and Gore faced considerable political pressure— internal as well as external. Congress would not pass a treaty that was too hard on the West. As a compromise, the United States administration asked developing nations to agree to emissions limits. In rebuttal, those countries argued the United States, Japan, European states, and other advanced countries, by far the biggest polluters, should shoulder most, if not all, of the burden.

After heated debate threatened to end the conference prematurely, an agreement was finally struck. The United States and other industrial nations agreed to stricter binding limits on six greenhouse gases, including carbon dioxide. The United States committed to seven percent cuts, Europe eight percent, and Japan six percent below 1990 levels, achieved within the next fifteen years. In return, they gained the right to engage in "emissions trading," paying low polluting countries for the right to use part of their quota. In other words, if New Zealand came in way under its limits, it could sell its extra portion to a more industrialized nation such as Germany. Then

Germany could exceed its pollution requirements, by the amount it purchased, and still fulfill its treaty obligations.

To some political pundits, particularly those supporting the Clinton administration, the Kyoto Protocol was a milestone. To others, such as representatives of Greenpeace, it was a farce; it didn't go far enough. To still others, such as many industrial leaders, it constituted needless surrender to the demands of the developing nations. Its true impact will not be felt for many decades; only then might historians draw proper conclusions.

Global warming is by no means the only environmental catastrophe challenging our world. Equally sinister are the atmospheric effects of artificial gaseous substances known as chlorofluorocarbons. These gases have not just contributed to the greenhouse effect, they have also depleted Earth's ozone layer, generating a crisis for all living beings on our planet.

Hopping Mad

Where have all the frogs gone? The swamps and streams have grown strangely quieter, bereft of their guttural symphony. Like the traditional role of canaries in pointing out the hazards of coal mines, the current perils of amphibians—not just frogs, but toads and salamanders as well, serve as harbingers of grave dangers to the human race. Around the world, survey upon survey has shown their numbers are declining and their deformity rates are growing. If nothing is done, there may eventually be none left.

In 1997, a team of biologists from Oregon State University published a critical study clearly linking amphibian deaths and abnormalities with an increase in the amount of ultraviolet-B (UVB) radiation permeating Earth's atmosphere. Headed by Andrew Blaustein, the group examined rates of embryonic death and deformity among the long-toed salamander living in lakes of the Cascade Mountain Range, their natural habitat for thousands of years. These creatures reproduce by laying eggs in shallow water, making their embryos especially sensitive to radiation levels.

Blaustein's team compared two sets of salamander eggs. The first was exposed to natural sunlight in their native habitat. The second was shielded by a special filter, akin to a pair of good sunglasses, that blocked all UVB light. In the former case, 85 percent of the embryos died. Of the 15 percent surviving, all but four of the animals were deformed (with missing or extra appendages). The latter (UVB protected) group fared considerably better by all statistical standards; 98 percent survived and virtually none were born deformed. Since the only difference in their conditions was their exposure to Earth's atmospheric UVB, only that factor might explain the contrast between the two populations. Thus Blaustein and his colleagues concluded solar UVB caused the deaths and deformities. By extension, they surmised the problems faced by other amphibian species were due to the same reason.

UVB light is one of three types of ultraviolet radiation produced by the Sun. The other two kinds are called ultraviolet-A (UVA) and ultraviolet-C (UVC). Ultraviolet light has a slightly shorter wavelength than visible light; therefore it cannot be seen, but only felt. The shorter its wavelength, the greater its energy, and the more harm it might render to living creatures. UVA light is the least energetic of the three types. Though much of it reaches Earth's surface, it is believed harmless. With somewhat shorter wavelength, UVB radiation is powerful enough to cause damage, particularly to the sensitive coverings of living organisms (eyes, skin, embryonic tissues, etc.). Unfiltered, it causes ailments such as cataracts (clouding of the cornea that can be triggered by radiation) and skin cancer (radiation-induced alterations in the genetic material of skin cells). Even more harmful to plants, animals, and people is UVC light, which can be lethal even in small doses, causing radiation poisoning.

Fortunate to man, woman, and beast, Earth is surrounded by a protective shield that blocks most of the UVB radiation and almost all of the UVC rays: the ozone layer. Ozone is an unstable combination of three oxygen atoms (the ordinary molecular oxygen we breathe has only two) that is especially efficient in absorbing ultraviolet light. High up in stratosphere a dilute layer of that reactive substance—so thin that if it were somehow spread over the ground it would have a thickness less than 1/8 of an inch—is responsible for maintaining the health of life on Earth.

Why then is UVB radiation on the rise? Isn't the ozone layer doing its job? Lamentably, scientists have known with certainty since the mid-1980s the upper atmospheric ozone has become increasing dilute, especially over the colder parts of the globe. The thinning out of the Earth's ozone layer, and the subsequent increase in dangerous ultraviolet solar radiation, constitutes an environmental crisis of monumental proportions, comparable only to that of the greenhouse effect.

The causes of ozone depletion are now well known. In 1928, a group of researchers at General Motors, led by chemist Thomas Midgley, invented an inert (nonreactive) type of gas, a combination of chlorine, fluorine, and carbon atoms, which they called a CFC. Over time, a number of similar chemicals were developed, which found uses as refrigerator coolants and spray can propellents. Because they were odorless and nontoxic, they represented an advance over the corrosive substances, such as ammonia and sulfur dioxide, that were used before for refrigeration. Eventually, as marketed by du Pont under the trade name Freon, CFC use in refrigerators and spray cans became ubiquitous.

In the early 1970s, independent British scientist James Lovelock—best known for his formulation of the Gaia hypothesis that the Earth is like a biological organism—was curious to know how much of these CFCs were retained by the atmosphere and how much blew off into space. He thought these chemically inert gases would serve well as tracers. That is, by staying intact and swirling around with the rest of the atmosphere, like buoys in a whirlpool, they might provide a guide to how air circulates. Lovelock did some atmospheric testing and was astonished to find that the entire stock of CFC gases released since the 1930s remained in the air. Sticking to the atmosphere like gum to the bottom of a schoolroom chair, these substances appeared to linger unaltered for decades and decades.

Several years later, Sherwood F. Rowland of the University of California at Irvine along with Mario Molina, currently at the Jet Propulsion Laboratory in Pasadena, California, decided to investigate what might become of the CFCs in the atmosphere. After performing a series of calculations to determine the ultimate fate of those chemicals, they were stunned. They determined each CFC molecule had the capacity to rise to the stratosphere and eat up tens of thousands of ozone molecules, one after the other like a many course meal. The combined appetites of all of the CFCs released by human activity to date would be sufficient to devour at least a few percent of the ozone layer, enough to increase significantly the world's exposure to harmful ultraviolet radiation.

CFCs, they found, destroy ozone in a chain reaction. As each molecule reaches the stratosphere, ultraviolet radiation breaks off its chlorine atom. The chlorine then reacts with ozone to form a compound, releasing molecular (breathable) oxygen in the process. The compound, in turn, reacts with free oxygen atoms, to create more molecular oxygen, freeing up the chlorine atom again in the process. Now liberated, the chlorine can find another ozone molecule with which it can perform its mischief all over again. For a single chlorine atom rising through the atmosphere, this process can recur thousands of times.

Naturally, Rowland and Molina's paper caused a considerable stir. In the late 1970s and early 1980s, atmospheric scientists around the world urged immediate steps be taken to ban CFCs. They counseled that, for the time being, spray cans should be used as little as possible. Though the public complied with these requests for the most part—switching to roll-on deodorants and cutting back on air-conditioning to do their part for humanity—the chemical industry was reluctant to eliminate CFCs until good substitutes could be found. Besides, industry leaders argued, the matter still had not been decided; more tests were necessary before radical steps such as banning should be taken.

In 1985, scientists found all the evidence they needed in the skies above Antarctica. Researchers from the British Antarctic survey, using NASA's Nimbus 7 satellite, detected a mammoth hole in the ozone layer over that frozen continent, an area where more than half the ozone content was being lost each spring. Records showed the depleted region would gradually recover each fall, only to balloon again when the warm sun returned to the southern hemisphere. Moreover, the problem seemed to be getting worse over time.

Computer analysis revealed the source of the great hole. These models indicated microscopic ice crystals in the skies over Antarctica

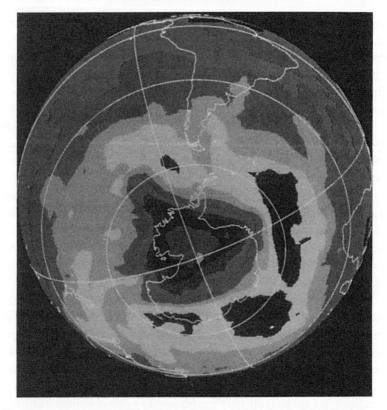

The ozone hole over Antarctica, as depicted with data from the Nimbus-7/Total Ozone Mapping Spectrometer (TOMS) instrument. In late November 1992, when this image was taken, the size of the hole was 1.7 million square miles. (Courtesy of NASA.)

were trapping chlorine atoms from CFCs during the long cold winter, and releasing them again during the spring thaw to do their damage. No wonder as the sun rose over Antarctica each springtime, hordes of chemical ozone-eaters conducted their annual banquet.

The discovery of the Antarctic ozone hole, followed in 1987 by the finding of a second hole over the North Pole, shocked the world into action. Several international treaties were signed to phase out the use of CFCs and other chemicals that release chlorine into the stratosphere. The most well known of these was the Montreal Protocol, signed by forty-five nations in 1988. That treaty, which called for a 50 percent cut in production of ozone-depleting chemicals, was quickly rendered obsolete by new agreements in 1989 calling for a full global ban by the year 2000.

Before we breathe a sigh of relief that these destructive chemicals are finally on the way out, let us consider this sobering fact: Even after production stops, most of the damage has yet to be done. CFCs tarry for many decades before they are rendered inactive. Sadly, they are expected to destroy far more of the ozone layer before their effects are muted. Only then might the ozone layer return to its original state and offer full protection against the deleterious effects of UV light. In the interim, years of heightened skin cancer and cataract rates for humans, as well as increased deformity rates for animals, will serve as painful testimony to the irreversible damage of pollution.

Reviving Our Planet

In the past century the human capacity for self-destruction has grown immeasurably. Our ability to destroy our world has manifested itself in countless ways—epitomized by the sad story of Hiroshima and Nagasaki, but also symbolized by more subtle tales such as the record of ozone depletion in the skies over Antarctica.

Nuclear catastrophe represents the ultimate way in which humankind might devastate its mother planet, perhaps irreparably. Let us hope that its mad potential remains a warped thought and never produces a ruined reality. However it is far from the only method by which the follies of our race might wreak havoc upon our world.

Environmental pollution, particularly through disruptions of Earth's delicate atmospheric balance, might bring on a more gradual form of doom—a slow strangulation of the world's living habitats. In the long term, the unchecked proliferation of greenhouse gases and ozone-depleting chemicals would make Earth a far less hospitable place to live. Fortunately, through global treaties, some steps have been taken to curb these releases. Are these efforts significant, or are they purely cosmetic—mere lip service for the sake of appeasing a frightened public? Let's hope for the former, but only time will tell.

Scientific monitoring of atmospheric conditions—average global temperature and levels of ultraviolet radiation will let us know our progress. As in the case of a patient experiencing distress, if Earth's situation grows worse, much more draconian measures will need to be taken—maybe even within the next few years. In that event, one hopes reason will win out over politics, and nations will embark upon whatever course necessary to rescue our planet. Otherwise our children's children will rightly curse what nightmares our generation has wrought: deadly sunlight, flooding seas, poisoned air, ruined farmlands, and other calamities.

Unfortunately, while striving to avoid nuclear armageddon and environmental devastation, our civilization could be ravaged by another kind of disaster, from beyond, rather than from within. We are denizens of the cosmos and therefore subject to the unpredictable nature of cosmic catastrophes. Though Earth today is teeming with life, an unforeseen astronomical event—a cometary impact, asteroidal bombardment, or interstellar collision—could suddenly wipe out the bulk of all living species. Unless we master the technology to divert such objects, or the ability to flee into space and settle on other planets, if a large enough space rock drops on Earth, the human race might be doomed.

Even if our world is lucky enough in the future to avoid lethal impact with a celestial encroacher, its lifetime would still be limited. Billions of years from now, the Sun will swell up into a bloated fireball, and then shrink down to a cold, dim orb. Lacking the sustaining heat of its parent star, Earth would no longer be habitable. A frozen remnant of its glory days as a planet, it would be as lifeless as Pluto is today.

Or, even before the Sun dies, the explosion of one of our stellar neighbors—in a sudden outburst known as a supernova—could fatally disrupt our system. Depending upon the stability of nearby stars, such lethal blasts could happen at any time. Traveling at the speed of light, the energy from a proximate supernova explosion could reach and possibly destroy the Earth's environment.

Finally, interstellar travel might provide temporary respite from such unfortunate events by allowing us to establish distant new alcoves for the human race but eventually, the universe itself will meet its demise. At that point, all life on the multitude of planets revolving around the myriads of stars that speckle space will be extinguished.

The horsemen of the apocalypse might be outrun for countless eons, but they cannot be outraced forever.

We will next discuss ways in which astronomical events will someday seal the doom of our species, and, eventually, the fate of all living beings that may exist in the cosmos. We consider three possible doomsday scenarios: fatal impact with an extraterrestrial body, the death of the Sun, and the demise of space itself. Of these possibilities, the first could conceivably be perpetually avoided, and the second and third are sadly inevitable. Though it has clicked out a steady rhythm for billions of years, the great clock of the cosmos must someday cease ticking.

PART III

FATAL EQUATIONS

COSMIC APOCALYPSE

CHAPTER SEVEN

IN THE DAYS OF THE COMET

*I glanced up, and behold! The sky was streaked with bright
green trails. They radiated from a point halfway between
the western horizon and the zenith, and within the
shining clouds of the meteor a streaming movement had
begun . . . with a crackling sound, as though the whole
heaven was stippled over with phantom pistol shots . . . I
stood for a moment dazed, and more than a little dizzy. I
had a curious moment of purely speculative thought.
Suppose, after all, those fanatics were right and the world
was coming to an end!*

—H. G. WELLS, *"In the Days of the Comet"*

From Halley to Hale–Bopp

On the night of July 22, 1995, two American astronomers independently noticed a faint new object in the skies. Within minutes of each other, Alan Hale, of the Southwest Institute for Space Research in Cloudcroft, New Mexico, and Thomas Bopp, an amateur astronomer in Phoenix, Arizona, each used their telescopes to spot a minute, fuzzy image in the celestial dome where none had been seen before. They received joint credit for discovering

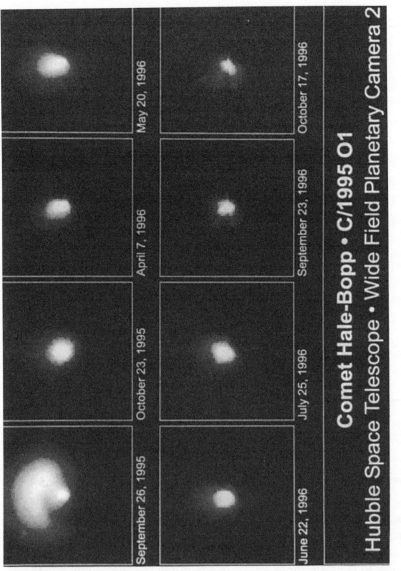

Comet Hale-Bopp • C/1995 O1

Hubble Space Telescope • Wide Field Planetary Camera 2

September 26, 1995 | October 23, 1995 | April 7, 1996 | May 20, 1996

June 22, 1996 | July 25, 1996 | September 23, 1996 | October 17, 1996

The progression of Comet Hale-Bopp toward the Sun from 1995 to 1997. Notice how the comet's nucleus changes as it is warmed by the Sun. (Courtesy of NASA.)

a new twenty-five-mile diameter comet, Hale–Bopp, fully twenty months before it experienced its closest approach to Earth.

By March 1997, Comet Hale–Bopp had traveled over a billion miles toward the inner Solar System since its first telescopic appearance, and had become noticeably visible with the naked eye. Its wispy double tail, with blue and white branches, stretched out approximately 50–60 million miles across space, nearly two-thirds of Earth's distance from the Sun, made it an eerie, otherworldly addition to the nocturnal tapestry.

Although Hale–Bopp never posed a true danger to Earth, its presence had a profound psychological impact on members of the Heaven's Gate group and other believers in pseudoscience. Its closest approach to our planet was 122 million miles, a comfortable distance away. Yet as it stretched out its wings like a glowing raven flying high in the cool night air, it became seen by these anxious observers as a portent of doom. This association was amplified by the near-coincidence of the coming of the comet and the turn of the millennium. Assorted self-proclaimed prophets, posting their messages on the Internet, anonymously or otherwise, quoted passages from Revelation to predict that Hale–Bopp would wreak havoc upon Earth, through triggering volcanic eruptions worldwide, and bring about armageddon. The Heaven's Gate cult believed an alien spaceship lurked behind the comet, using it as cover for a close approach to Earth. In swallowing poison, cult members believed they were transporting themselves to that craft.

When Chuck Shramek, an amateur astronomer, published a photograph showing a bright speck behind the comet that he could not identify, discussion groups on the Internet became forums for impassioned believers in government conspiracies that were seen as hiding the "truth" about Hale–Bopp. Alan Hale tried to clear up the matter with his public statements, demonstrating that the photographic "blip" was merely a bright background star, and could not possibly be an alien spacecraft. Though Hale was manifestly correct, his efforts did not convince Shramek's supporters; they accused him of being part of the "cover up" as well.

Historically, there have been many other incidents of panic upon the arrival of a comet. For example, in the opening decade of this

century, scientists' detection of cyanogen (a poisonous gas when in-
haled in high enough concentrations) in the tail of Halley's comet set
off a short-lived scare. In 1910, as that famous celestial snowball approached our planet on one of its periodic journeys around the Solar
System, panicked individuals around the United States, misinter-
preting data disseminated by the Yerkes Observatory of the University of Chicago, put forth theories as to what would happen if Earth
passed through its wispy posterior region. One amateur chemist,
who knew little about comets, predicted the cyanogen would combine with atmospheric oxygen to form nitrous oxide.[1] Once Earth's
air turned into "laughing gas," he purported, humanity would
asphyxiate as it giggled itself to death. Members of the scientific
community, including Yerkes astronomers, issued statements of re-
assurance, proving that the amount of cyanogen in Halley's tail
would be minuscule—too small to have any discernable effect. Still,
the crisis didn't resolve itself until Halley cleared Earth by a safe
margin and returned to deep space.

The doomsday panic of 1843 and 1844, coinciding with a membership boom in the apocalyptic Millerite movement, was similarly
precipitated by fear of a comet. In that case, the comet was not close
at all; rather, it was extraordinarily brilliant and dragged along a gargantuan tail. Because Miller predicted the world would end that
year, the comet was considered to be a bad omen. Many frightened
Christians viewed it as one of the portents of doom mentioned in
Revelation: the great star "Wormwood" crashing down upon Earth.

Though these reactions were extreme, it is natural to have a certain degree of apprehension about cometary arrival. The passage of a
comet through the heavens is an unmistakable symbol of the precarious nature of our existence on this planet. There is something about
the way these objects appear mysteriously in the heavens, creep
stealthfully toward us, and, very occasionally, pass so close to the
Earth or, on rare occasion, collide with our planet that reminds us of
our species' mortality. Asteroids, in spite of their less foreboding appearance—creating no visible streak across the sky—nevertheless represent another potential threat. For if an Earth–comet or Earth–asteroid
collision were to occur with enough striking power (estimated by researchers David Morrison, Clark Chapman, and Paul Slovic to be im-

pacts with explosive yields of hundreds of millions of megatons or greater[2]) the human race would surely be doomed. With odds per year of less than one in 100 million, such an unlikely crash could obliterate advanced life on our planet. Enough dust and debris would be kicked up by such a collision to blot out the Sun for many months, perhaps even years. Temperatures would drop dramatically, in a sharp cooling effect similar to nuclear winter. Crops would fail and animals would starve. Robbed of the sun's energy and deprived of its main food sources for such a prolonged period, the human race would likely die out. In essence, we would meet the fate of the dinosaurs.

To understand how the impact of a large comet or asteroid could possibly represent doom for our planet, let us examine the origins and properties of these potential intruders.

Rocky and Icy Intruders

Billions of years ago, the Solar System resembled a whirling dust storm. Enormous clouds of hydrogen and helium gas swept tiny grains of matter around and around in rapid rotation. As the dust ball twirled, it began to flatten along a single plane. Gradually, it shaped itself into the form of a pancake, evolving into an object known as a protoplanetary disk.

As time went on, the center of the original dust ball condensed under the weight of its own gravity. Once it reached a certain critical mass, its hydrogen atoms began to fuse together in a thermonuclear reaction. Soon it began to glow from the energy produced in its fusion process. This fiery center—which we identify as the Sun—continued to grow hotter and hotter, bathing nearby regions with its radiation.

Meanwhile, the outer parts of the disk undertook a different sort of evolution. Over time, numerous grains of dust collided with each other and began to stick together under mutual gravitational attraction. As this process continued, larger and larger clumps of matter were fashioned. Astronomers refer to these rocky chunks as "planetesimals."

In the next stage of the evolution of the Solar System, four distinct regions were formed. In each of these zones, disparate behaviors took place, depending on the region's proximity to the Sun. In

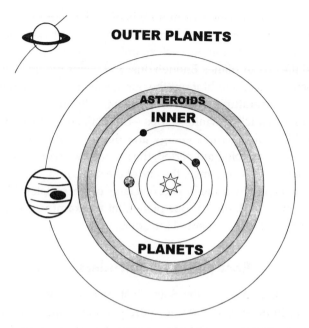

The major regions of the Solar System (not to scale). These include the inner plane-tary zone, the asteroid belt, the outer planetary zone, and the Kuiper belt (beyond the range of the depiction).

the inner planetary zone, the domain closest to the Sun, rocky plan-etesimals bashed together repeatedly through random collisions. Gradually, they merged with each other to form bigger and bigger bodies. Four of the largest "rocks" soon became massive enough, and thus gravitationally powerful enough, to attract other planetesi-mals to themselves. These formed the cores of what would come to be known as Mercury, Venus, Earth, and Mars. Over time, these four inner planets acted as celestial vacuum cleaners, sucking in inter-planetary debris from their surroundings, and emptying most of nearby space. Eventually, the inner zone of the Solar System was, for the most part, free of planetesimals and only the four planets (and several moons) remained.

Let's now skip to the third part of the Solar System, the do-main that presently encompasses the five outer planets, Jupiter,

Saturn, Uranus, Neptune, and Pluto. Because this region is much farther away from the Sun than the inner planetary zone, it is naturally much colder. Therefore, many substances that exist in the inner Solar System as liquids or gases, can be found in the outer reaches exclusively as solids. Common compounds such as methane, ammonia, and water are present in the outer Solar System only as ices.

Like the inner planetary zone, the outer planetary zone went through a similar period of condensation. Over time, smaller particles in that region came together and formed progressively larger bodies. However, in this colder part of space, most of the material that coagulated consisted of frozen chemicals, rather than rock. Rocky planetesimals formed the mere seeds for thick outer shells of ice. Gradually, these icy bodies merged into what we know as the outer planets. Ultimately, most of the remaining material in this region became captured by these five planets' strong gravitational fields.

That leaves two regions of the Solar System for consideration. One of them, the second zone from the Sun in our accounting, is the area known as the asteroid belt. There, in the band of space between Mars and Jupiter, exist rocky planetesimals that have never mustered enough gravitational attraction to unite into planets. In this zone, the gravitational forces of Jupiter and the Sun maintain a constant tug of war, preventing larger bodies from forming. Therefore, the rocky objects—the asteroids—that exist in this region remain forever in their primordial state.

The bulk of asteroids track steady orbits around the Sun, while staying within the rough boundaries of the asteroid belt. Occasionally, though, an asteroid becomes thrown off course and ends up within the region of Earth's orbit. These rogue objects, called NEAs (Near-Earth asteroids), constitute the greatest threat for Earth–asteroid collisions.

Aside from the asteroid belt, there is another part of the Solar System where planets never formed—the enormous, ball-shaped region known as the Oort Cloud. This remote sector completely surrounds the flattened disk on which the planets enact their orbits. Hundreds of billions of miles from the Sun, the Oort Cloud is un-

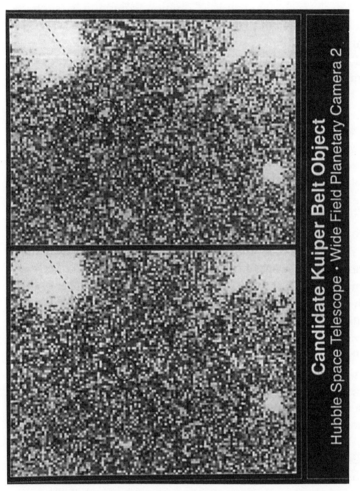

Candidate Kuiper Belt Object
Hubble Space Telescope · Wide Field Planetary Camera 2

Sample data taken by the Hubble Space Telescope showing one of the myriads of icy objects that form the Kuiper belt. The Kuiper belt is so distant, and its objects so small, that even the powerful Hubble cannot image them very clearly. (Courtesy of NASA.)

doubtedly the coldest sector of the Solar System. Floating within its frigid domain are trillions of chunks of frozen chemicals and dust. These are remnants of the icy planetesimals that once occupied most of the outer Solar System, from Jupiter on out. These celestial snowballs harbor substantial quantities of water, methane, ammonia, carbon dioxide, and other simple compounds, locked up in their frozen states.

The closest part of this zone of frozen chunks is a flattened region, called the Kuiper belt, lying just beyond the orbits of Neptune and Pluto. The Kuiper belt, named for astronomer Gerard Kuiper who predicted it in 1951, is near enough to Earth that some of its small, icy denizens can be viewed with the Hubble Space Telescope. There have been a few reports of observers who viewed the Kuiper belt while using a less powerful telescope. It is estimated that the Kuiper belt harbors tens of thousands of object larger than fifty miles across, and many more even smaller.

Occasionally, frozen chunks depart the Kuiper belt—as well as the more distant reaches of the Oort Cloud—jostled out of orbit by the gravitational influence of a passing body. Once disturbed, they either leave the Solar System, or begin the long voyage toward the Sun. As they approach the warmer region of space in the vicinity of the Sun, their ice begins to vaporize. These vapors form an extensive trail that lags behind the chunk as it moves. From Earth, we view such a system as a comet streaking through the skies. We observe the original ice chunk as the comet's nucleus, and the vaporous exhaust as the comet's tail.

In summary, the Solar System is composed of a number of diverse regions. Several of these zones, namely the inner planetary region and the asteroid belt, are warm enough to harbor rocky structures. Sectors farther from the Sun, such as the region beyond Jupiter, favor icy formations. Asteroids and comets are the remnants of the original materials that once occupied the entire Solar System. Normally, they reside in their own parts of space—the asteroid belt, Kuiper belt, and the Oort Cloud respectively. Occasionally, they are drawn away from their native habitats and flung toward the inner Solar System, sometimes assuming trajectories that cross Earth's. On

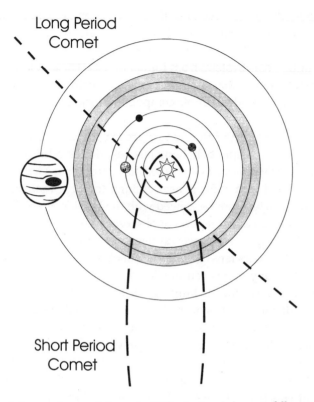

Short period versus long period comets. While short period comets follow regular el-liptical orbits around the center of the Solar System, long period comets, pulled in-ward by a disturbance of the Oort cloud, engage in far more erratic trajectories.

extremely rare incidents these objects crash into Earth, explaining, perhaps, many of the great extinctions.

Short period comets are those that appear regularly. They originate in the Kuiper belt and follow simple periodic orbits around the Sun. They may voyage around the Sun along elliptical paths for many hundreds of years. Halley's comet is a good example of a comet with more or less clockwork behavior; it traverses its orbit roughly every seventy-six years.

In contrast, long period comets seem to arrive like a bolt out of the blue. They stem from the deepest reaches of the Oort cloud, and have generally been seen only once. The mechanism by which long period comets are wrested from their original positions in the Oort Cloud is not yet fully understood. Though part of the process is clearly due to chance gravitational tugs or collisions that alter the courses of these objects, it is not known if there is a systematic component as well. Could there be something in space that regularly hurls comets toward our region?

Stars of Doom

As we've seen, many scientists today believe that regular encounters with asteroids and/or comets have been responsible for the puzzling episodes of mass extinction present in the fossil record. For example, following the hypothesis advanced by Luis Alvarez and his colleagues in the 1970s and 1980s, the currently favored view of why the dinosaurs died out is that the Earth suffered from the impact of an extraterrestrial body.

In 1983, examining the past 250 million years of evolutionary history, as documented in the fossil record, paleontologists David Raup and J. John Sepkoski of the University of Chicago found a curious regularity to periods of great loss of species.[3] They extended the extraterrestial impact theory by proposing that extinction events have occurred periodically, about once every thirty million years. They detailed nine catastrophic events that fell into this pattern, and found only two major gaps in sequencing. Astronomers have yet to discover the cause of such a pattern, if it indeed is genuine.

One possible explanation for these apparent thirty-million-year cycles is called the Nemesis hypothesis. In 1984, astronomers Marc Davis and Richard A. Muller of the University of California, along with Piet Hut of Princeton University, proposed that an invisible, distant companion to the Sun, called Nemesis, periodically disturbs the Oort Cloud. According to this theory, Nemesis, a very dimly shining star, orbits the Sun once every twenty-five to thirty million years or so. Each time it passes through the heavily populated regions of the Oort Cloud, its gravitational force dislodges comets from their positions. These comets, once disturbed, come soaring toward the vicinity of Earth.

According to the Nemesis hypothesis, these periodic comet showers caused the cyclic mass extinction of species on Earth. The dinosaurs, for example, were wiped out during one of these episodes. And in fifteen to twenty million years from now—so the theory goes—Nemesis will move through the Oort Cloud again, releasing another death volley of comets and hurling them toward our doomed planet.

Rightly so, most scientists have rejected the Nemesis hypothesis. There are many reasons why this theory is untenable. First of all, it is absurd, at this point, to imagine the Sun has a hidden companion. Light recording instruments, from the IRAS infrared satellite to the Hubble Space Telescope, have probed the Solar System and found no trace of another star. True, some types of stars, such as red dwarfs (stars that have barely enough nuclear fuel to shine), burn dimly. However the current crop of telescopes possess the sensitivity required to detect such a fading star, if it were actually out there.

Moreover, a host of geological evidence indicates most of the craters believed to be linked with mass extinction events are asteroid impact basins rather than comet crash sites. The Chicxulub crater, for instance, has a structure that seems to indicate it was generated by a large asteroid.

From what we know of asteroids and comets, this is not surprising. Comets, composed mainly of ice rather than of rock, are much more fragile than asteroids. Unless they harbor hard centers, they are less likely than asteroids to form visible craters upon impact.

Comet expert Fred Whipple, of the Smithsonian Astrophysical Observatory, sums up well, in his book *The Mystery of Comets*, these widely-held misgivings about both the Nemesis hypothesis and the theory that periodic comet showers caused mass extinctions:

> With regard to the great extinctions, my bet would be at least even money favoring each of the following wagers: *Some* of the major extinctions were caused by great impacts; neither great interstellar molecular clouds nor a Nemesis companion to the Sun triggered a pseudoperiodicity in the extinctions; comets produced few impact craters on the Earth, if any, and no extinctions, except possibly for some

extremely large and old comets whose cores were heated to become superficially indistinguishable from asteroids.[4]

Though the Nemesis hypothesis seems untenable—there is no reason to think the Sun has a dim companion—some researchers believe that distant stellar objects occasionally do approach our system and cause cometary disruptions. At the 1997 Hipparcos symposium in Venice, a team headed by Robert Preston and Joan Garcia Sanchez of the Jet Propulsion Laboratory in Pasadena, California, predicted that, over the next million years, eight stars will approach within five light years (thirty trillion miles) of the Sun. (Currently the Alpha Centauri system, including Proxima Centauri, the nearest stellar object, is the only set of stars that close.) As they near the Solar System, these stars could gravitationally affect objects in the Oort Cloud enough to alter their courses and send them plunging toward Earth. In just ten thousand years, for instance, Barnard's Star, a fast moving red dwarf nicknamed the "Greyhound of the Skies," will zoom by the Sun less than four light years (twenty-four trillion miles) away. It will then be the closest star to Earth, besides the Sun, and thus exert a formidable influence on the outer reaches of the Solar System.

The Jet Propulsion Lab team conducted their study by first observing stars with the European Space Agency's highly precise Hipparcos satellite, and then calculating the velocities of those that seemed headed most directly toward the Sun. The researchers determined the members of this group by picking out the stars with the least amount of sideways motion. They reasoned that if an object is hardly moving sideways with respect to us, it must either be coming toward us or heading away from us.

To determine the speeds of approaching and retreating stars, the group used a common astronomical technique known as the Doppler method. This approach relies on the fact that the spectral lines of bodies moving away from us are shifted toward the red end, and of those coming toward us, to the blue end. For example, the red lines of an approaching star's characteristic hydrogen spectrum would seem more orange, and the orange lines more yellow. The faster an object approaches or recedes, the more noticeable the effect.

The Doppler effect is most familiar when applied to sound. The high pitched wailing of the police car's siren as it approaches, and the gradually deepening of the sound as it speeds away are common street noises. These changes in pitch stem from wavelength shifts in the sound waves produced by the siren. Similar Doppler shifts take place in the light waves of fast-moving celestial objects, such as stars.

By measuring their current positions and speeds, the team plotted out the trajectories of hundreds of stars that appeared to be headed in the general direction of the Solar System. Of these, they noted eight they thought would pass close enough to create cometary disturbances.

One star observed by Preston and Sanchez seemed to have an especially ominous path: a direct collision course toward the Oort Cloud. According to their calculations, a red dwarf, called Gleise 710, more than one hundred thousand times heavier than Earth, is heading right for the Solar System at almost ten miles per second. Currently, Gleise 710 lies sixty-three light years away in the constellation Ophiuchus. In about one million years, it will likely collide with the Oort Cloud head on, precipitating a massive storm of comets raining down on the inner Solar System. Under these circumstances, Earth may very well be pelted with unwanted, and possibly lethal, giant "dirty snowballs" from above. (The term, "dirty snowball," an apt description of a comet as a mixture of ice, grains, and fine dust, was coined by Fred Whipple.)

The Great Melt

What are the chances asteroids or comets will hit Earth in the near future? Under what circumstances might they do significant damage to our planet and even endanger our race? Generally speaking, as Morrison, Chapman, and Slovic point out in their comprehensive study, "The Impact Hazard,"[5] published in *Hazards Due to Comets and Asteroids,* the larger the asteroid or comet, the more rare the impact, but the greater the harm.

The mechanism by which a comet or asteroid would wreak damage is fairly well understood. Entering Earth's atmosphere, an astral intruder would generate an enormous cushion of air beneath it. Tremendous air pressure would eventually decelerate the intruder, and convert its energy into a powerful shock wave.

The type of destruction caused by the blast would depend on its locale. Scenarios would vary if it is above land or over the ocean. If the shock wave were to detonate above land, it would scoop up the ground and eject massive amounts of soil high up into the atmosphere. The region from where the soil is displaced would become a crater—its size proportional to the strength of the blast.

Not just dirt would be funneled up into the air. The hefty winds generated by the outburst could hurl trees, street signs, lampposts, pieces of buildings, and other debris high up into the sky. Broken glass from shattered windows could come whirling through the air, injuring those unlucky enough to be in the area. Houses could collapse in on the unfortunate families living in the impacted region. Like the aftermath of a bomb blast, whole forests could be felled, entire villages wiped off the face of the map.

Upon the ocean, a strong enough impact would generate tsunamis (tidal waves): formidable walls of water that would dwarf ordinary waves. If the object that hits is a comet, melted ice from its exterior would rain into the ocean, causing water levels to rise even further. Gargantuan floods produced by the collision could destroy coastal regions and decimate island communities.

The tsunamis generated by the crash of an extraterrestrial body into the ocean would likely be much higher, and vastly more destructive, than even the tallest, deadliest waves on record. Tidal waves are generally caused by strong earthquakes that rumble less than thirty miles beneath the floor of the sea. (In spite of their name, they are never produced by the tides.) They can also be created during powerful volcanic eruptions—such as the explosion of Krakatoa, discussed earlier. Krakatoa's 1883 eruption produced waves as high as 115 feet, killing more than thirty-six thousand people. They were not the deadliest tsunamis of all time, however. The record-holder belongs to a tidal wave that occurred in 1703 near Awa, Japan, that killed more than one hundred thousand people. The waves produced by extraterrestrial impact, though, could easily break these records—rising hundreds of feet high and killing many hundreds of thousands, or even millions.

One might wonder what would become of the remnants of the crashing object. Unless it had a hard rocky core, a comet would likely be fully vaporized by the time it would hit the Earth. Its ice would turn completely into water vapor. In the case of an asteroid, on the other hand, whatever is left of the original body would plummet to the ground as a meteorite. Meteorites are falling objects from space, composed mainly of iron. They are to be distinguished from meteors: tiny bodies that whiz harmlessly through the upper atmosphere, causing brilliant displays as they disintegrate.

Depending upon the size of the encroaching comet or asteroid and the height at which the blast of impact is produced, the amount of damage caused by collision would vary. According to Morrison and his colleagues, objects greater than 1000 feet wide, colliding with Earth less than once every hundred thousand years, would wreak the most havoc. Generating blasts of greater than ten thousand megatons (million tons of TNT equivalent) and producing craters at least ten miles wide, these collisions would create certain widespread destruction—through floods, fires, and collapsed buildings—resulting in many thousands, or even millions, of deaths. Dust would fill the skies, block the Sun's warming rays, and temporarily lower global temperatures, in a manner similar to the effects of large volcanic eruptions or nuclear winter.

Computer simulations support the viewpoint the shock of a large comet (thousands of feet in diameter)—let alone an immense one such as Hale–Bopp (tens of miles in diameter)—would decimate the ecology of our planet, at least temporarily. In 1997, David Crawford, a planetary scientist at Sandia National Laboratories in Albuquerque, New Mexico, calculated that the impact of a half-mile wide comet (one-fiftieth the girth of Hale–Bopp) would cause damage equivalent to that of fifteen million atomic bombs of Hiroshima's power.

Crawford programmed sophisticated three dimensional simulations of cometary collisions on a new supercomputer at his laboratory to measure their deadly forces. In his model, an incoming comet vaporizes directly above the Atlantic Ocean, causing a monumental rain of evaporated and melted ice. The shockwave of its impact stirs up an additional hundreds of billions of tons of water, leading to tsunamis as high as three hundred feet. All low-lying coastal areas around the Atlantic, from Florida to downtown Manhattan, would soon become deluged. Picture Miami Beach turning into an underwater monument to art-deco hotels, and Wall Street becoming an aquarium filled with soggy ticker tape. Even the best swimmer, caught in such a tsunami, would have no chance whatsoever to flee. The wave would arrive so fast and so high, he or she would be knocked unconscious and drowned. Eventually, after the great flood had retreated, the swimmer's lifeless body would wash up onto shore, joining the remains of other former coastal dwellers. Casualities would likely number in the hundreds of thousands, perhaps even in the millions.

Yet as Crawford has remarked, the human race would probably still survive: "You'd certainly have devastating consequences on a regional scale, though it wouldn't necessarily cause the end of humankind," Crawford reported in a recent account of his results. "But all the coastal cities would be threatened by such an event."[6]

Crawford estimates comets of the size he studied hit Earth about once every several hundred thousand years. The main effect of such bombardments (if over the ocean) would be global flooding of low-lying coastal areas, rather than climatic changes of the kind that wiped out the dinosaurs and other species. Events that lead to certain mass extinction, he feels, occur much more rarely, involving either very large comets or asteroids. (Morrison, Chapman, and Slovic have conjectured that impacts of greater than 300,000 megatons, taking place approximately once in a third of one million years would cause enough damage to wipe out over 25 percent of Earth's human population.)[7]

John Lewis, codirector of the NASA/University of Arizona Space Engineering Research Center, has revived the notion that both the biblical tale of the Great Flood and the Babylonian Epic of Gilgamesh could have been based on an actual deluge triggered by a cometary collision. He ponders if "the clock of human history [was] reset to zero by an event (or more than one) that devastated civilization."[8]

In Lewis' opinion, universal flood stories constitute exaggerated retellings of catastrophic events that changed world culture. He remarks how odd it is our ancestors suddenly made the switch from hunter–gatherer to agricultural civilizations some eight thousand to ten thousand years ago. Perhaps, he speculates, after a period of flooding new dry land emerged around the world, making agriculture more productive. Humankind then found it economically feasible to cultivate tracts of land for food, rather than relying on found materials. Eventually descriptions of this profound transition found its way, in distorted form, into biblical and mythological literature.

As discussed earlier, current geological evidence does not support the idea that one or more cometary collisions triggered a universal flood. Rock samples dating back to biblical times have not shown traces of widespread flooding in the Middle East, nor indications of impacting objects. Furthermore, even if flooding did occur, it would probably have affected only coastal regions, not landlocked plains, and certainly not hilly or mountainous regions. Thus, Lewis' notion that the bedrock of our culture was fashioned from the debris of ancient cosmic disasters—that comet crashes caused flooding which in turn stimulated agriculture—though an intriguing concept in theory, finds no scientific proof.

Perils from the Skies

Thanks to improved computational methods, such as Crawford's recent study, science is pressing ahead toward an increased understanding of comet and asteroid collisions. There are several ways astronomers today are trying to obtain better estimates of the likelihood of a cataclysmic impact.

One approach is to learn the exact locations and orbits of all small celestial bodies in the Solar System. In theory, armed with this knowledge, one would be able to predict when an object is fated to collide with us. However, in practice, the main drawback with this method is that there are myriads of comets in the Solar System—especially in the mammoth Oort Cloud. Most are so distant that not even the most optically advanced telescopes have a chance to record them all.

Furthermore, even if we knew the positions of every comet and asteroid in the Solar System, it would be virtually impossible to track all of them. The dynamics of these objects—taken as a whole—are highly complex, in many cases even chaotic. (Chaotic dynamics have the property that a small change in initial conditions leads to substantially different subsequent behavior.) Therefore, even the most powerful computers in the world today couldn't provide us with precise orbits of each of these bodies for all times. Undoubtedly, interbody collisions or gravitational interactions would occur that wouldn't be anticipated. Typically, we would have to wait until a distant object in the Oort Cloud broke away from the pack and began to head toward the inner Solar System before we could track it unequivocally. We would discover the comet as little as two years before it crossed Earth's orbit (or impacted with us, in the worst possible scenario).

The population of near-Earth asteroids and short-period comets (those that have already left the Oort Cloud, and are engaged in elliptical orbits around the Sun in the realm of the planets), collectively known as near-Earth objects (NEOs), in the Solar System is much smaller, far easier to track, and therefore understood substantially better. Since these are the bodies closest to us and most likely to strike Earth, astronomers find them most worthy of study. Experts think that the bulk of these will be successfully tracked by search groups such as Spacewatch. Though a few rogue asteroids and comets might slip through the net, within decades the overwhelming majority of NEOs will be well known.

Astronomers channel information about recent findings and further analysis of asteroids and comets to the Minor Planet Center, directed by Brian Marsden, where it is made available to the research community. From this data, impact hazard experts such as David Morrison might continue to revise their forecasts.

Based on the current catalogue of NEOs, scientists predict that about seventy large comets and asteroids will pass within ten million miles of Earth during the next two decades. Of these, only a handful of those big enough to be dangerous are expected to come within one million miles. From an astronomical perspective, that's about as close as a bee's wing is from its torso, but fortunately these passages will be sufficiently distant from our planet (assuming that scientific forecasts prove correct) to ensure no threat.

Another approach to estimating our imminent chances of catastrophe is to observe what has happened in the past, and then extrapolate toward the future. This is what scientific experts often do to make forecasts. By analyzing cosmic impacts that have occurred on Earth, as well as on other planets, scientists are best able to calculate their potential for destruction in the future. Moreover, by counting the number of visible craters, they try to guess reasonably the number of unseen ones. From this information, in turn, they hope to glean the knowledge needed for credible projections of impact rates.

For years geologists have collected data about terrestrial impact craters. Over a hundred such remnants of cosmic catastrophes have been catalogued and studied. Probably the best known of these is Meteor Crater in northern Arizona, famed for its spectacular appearance. Discovered in the 1880s, it stands nearly a mile in diameter—much smaller than the site in Yucatan, but, being on the surface, far more visible. Resembling a bite taken out of an apple, it offers stark contrast to the flat desert topography around it. Now a tourist site, visitors come from miles around to gawk at the eerie scar left from an

The Moon's cratered surface shows the long-term effects of astral collisions. (Courtesy of NASA.)

asteroid crash that is estimated to have occurred over fifty thousand years ago.

The bulk of the impact craters that have been discovered, however, are discernible only to the trained eye. Thousands of years (and in many cases millions of years) of erosion have taken their toll on most of Earth's collision sites. Often wind, water, ground shifts, and other wearing down factors, have reduced them to structures barely distinguishable from their surrounding landscapes. And in many cases, such as Chicxulub crater, they are buried hundreds of feet underground.

The Moon, our nearest neighbor, is certainly the best known extraterrestrial body. Its rugged terrain has been explored by astronauts on several missions. Because it has no atmosphere, it has been found to be an outstanding example of rampant cratering without erosion. Its pockmarked face bears the unmistakable visage of eons of strikes by smaller bodies. By counting the number of the Moon's impact sites, and adjusting for the relative sizes of the two worlds, scientists are able to approximate better how many craters would exist on Earth if it weren't for erosion. (Counting craters on Mars and Mercury has led to additional information about the typical rate of interbody collisions in the Solar System.) They have estimated that the great majority of terrestrial impact craters have yet to be discovered (or are so buried that they will never be found).

Crash on Jupiter

We needn't look far back in geological history to consider examples of catastrophic events in the Solar System. In March 1993, Gene and Carolyn Shoemaker, a husband and wife team of astronomers, along with amateur astronomer David Levy, discovered a string of comet fragments headed toward Jupiter. They spotted these fragments—remnants of a comet that was subsequently named Periodic Comet Shoemaker–Levy 9 1993e—while examining exposed photographic plates. These images were taken with the eighteen-inch Schmidt telescope high atop Mount Palomar in California.

This was an unusual sighting, because the comet was first seen only after it had already been pulverized by the mighty gravitational

David H. Levy, codiscoverer of Comet Shoemaker–Levy, at the Jarnac Observatory. (Courtesy of David Levy; photo by Wendee Wallach-Levy.)

forces of Jupiter. Consequently, its image appeared on the plates as a bar, rather than as a spot. The team was confused at first, until they realized exactly what they had observed. Carolyn Shoemaker recalls her befuddlement upon first viewing the photographs:

> I came across this very strange-looking object. I thought it had to be a comet, but it was the strangest comet I had ever seen because it was bar-shaped. I turned to the others and said: 'I don't know what this is. It looks like a squashed comet.' We were all sort of stunned.[9]

Within months after the discovery, information was gathered about the cometary fragments, and predictions were made as to the date of impact. By late 1993, astronomers determined the Jupiter crash would occur during a period of several days, centered on July 19, 1994. They calculated the collision would take place on the side of the giant planet that would be facing away from Earth at the time. Therefore, it was determined observers would have to wait until

Jupiter turned its impacted face toward Earth before they would see anything of interest.

The second week of July 1994 rolled around, and the world readied itself for the big event. Viewing instruments around the globe—from amateur, homemade six-inch devices to the majestic Hubble Space Telescope—were trained on Jupiter, waiting for it to reveal its bruised visage. When the crash did come on the dates predicted, its sensational effects surprised even the most maverick of pundits. Many experts had predicted the effects would be subtle and short-lived, seen only briefly with large telescopes. Not so. The colors produced were dazzling and long-lasting. A hodgepodge of bright and dark patches dotted the surface, visible for days. Also, for many months after the impact, a dark band remained clearly visible in the planet's southern hemisphere where none had been before. These streaks and blotches were the endproducts of massive plumes of uplifted debris and displaced gases, created when the fragments hit. As time went on, patterns swirled, occasionally merged, dissipated, and finally faded. By the end of the phenomenal sky-show,

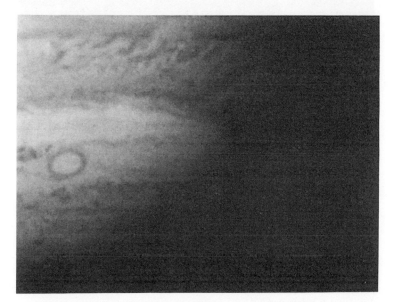

Jupiter, the largest planet of the Solar System. (Courtesy of NASA.)

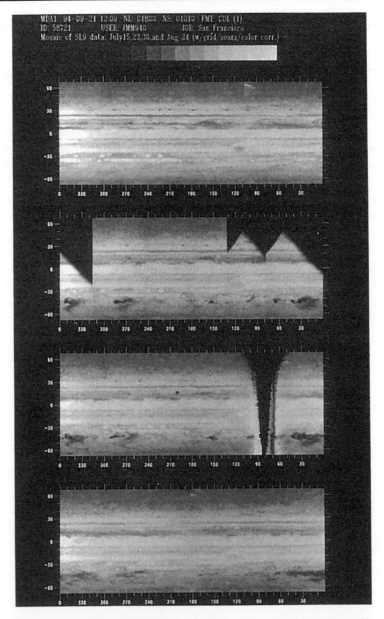

Impact sites on Jupiter resulting from the July 1994 collision of Comet Shoe-maker–Levy. (Courtesy of NASA.)

watchers agreed the crash of Shoemaker-Levy was the most spectacular astronomical event of the century.

After witnessing the scars Jupiter bore, one can only be thankful it wasn't our own planet that was struck. If the same comet had collided with Earth, the results would have been devastating. Any cities or towns situated near sites at major impact would have been demolished. Ocean levels would have risen, causing massive flooding of coastal communities. For quite some time, if the blasts were effective in stirring up enough dirt, the skies would have been full of soot and other debris. Consequently, our climate would have been disrupted for months, resulting in untold ecological ruin. Perhaps civilization itself wouldn't have survived the challenge posed by such enormous devastation.

We have been lucky that only a handful of extraterrestrial objects have crashed into Earth in the past few centuries. And fortunately, the occasional collisions that did occur were small ones, located far away from major population centers. For example, there is strong evidence a body of otherworldly origin hit Siberia in 1908. An enormous discharge, called the great Tunguska explosion, took place that year in a Siberian forest. A wooded region twenty miles in diameter was devastated by the blast, and thousands of reindeer were wiped out. Material hurled into the atmosphere by the explosion lit the skies of Eastern Europe for several nights. Luckily, because of the event's remoteness, nobody was killed. If it had been centered instead on a shopping district in Saint Petersburg or a bustling part of Moscow, thousands of Russian citizens would have perished in the fiery blaze. One can only hope that population centers are similarly spared the next time an asteroid or comet hits Earth.

Spacewatch

We cannot count forever on good fortune. Sooner or later, if civilization lasts long enough, our luck will run out and a massive fireball will demolish one of the teeming cities of our crowded planet, or perhaps even human life itself. Is there anything we can do to avoid such a tragedy, or at least to reduce its sorry consequences?

Astronomer Tom Gehrels is a lifelong optimist. As a youth, during the Nazi occupation of his native Holland, he did not flinch in the face of adversity. Rather, he bravely joined the Dutch resistance. In his current position at Spacewatch, he counsels we should meet impact hazards with a similar fighting spirit. In his opinion, we must

Tom Gehrels, founder of Spacewatch. (Courtesy of Tom Gehrels.)

develop a powerful enough system to anticipate and prevent such catastrophes. Then, well protected from their rare capacity for destruction, we can appreciate comets and asteroids for their natural wonder.

After the war, anxious to put the tragedy of the occupation out of his mind, and bubbling with enthusiasm about astronomy, Gehrels became a student at Leiden University, where he took classes with Professor Jan Oort. Oort, one of the greatest astronomers of the twentieth century, discovered the Oort Cloud. While Oort was rewriting astronomy's view of the origin and role of minor objects in the Solar System, Gehrels felt highly privileged to be studying under such a remarkable researcher.

After leaving Leiden, Gehrels found a position at the University of Chicago. There he developed a keen interest in tracking asteroids and comets, reveling in the thrill of the chase. As he reports in his autobiographical account, *On the Glassy Sea: An Astronomer's Journey:*

> Studying them involves some of the excitement of a hunt . . . Comets and asteroids have to be chased and carefully followed, or one loses them. Once the observer has pinned them down carefully, their orbital parameters are computed and published so they can be observed and studied again later.[10]

In the 1960s, he established wide recognition as an asteroid hunter, with the Palomar–Leiden survey. Gehrels ended up at the Lunar and Planetary Laboratory of the University of Arizona, where, in the 1980s, he established the Spacewatch project. Designed for the discovery and statistical study of asteroids and comets that lie in the Solar System, Spacewatch pioneered the use of ultra-sensitive electronic CCD (charge-coupled device) detectors for imaging. Replacing conventional photography, electronic detection allows images to be rapidly analyzed and, if of interest, transmitted directly to Brian Marsden's central clearinghouse for astronomical data at the Minor Planet Center in Cambridge, Massachusetts. There, Marsden's team plots orbits of comets and asteroids, and calculates their future positions. If a rogue asteroid or comet were fated to collide with Earth, Marsden's center would likely be the first to know—months or even years in advance.

Now directed by Dr. Robert McMillan of the University of Arizona, the Spacewatch group conducts its surveys with the 0.9 meter (three-foot diameter) Spacewatch telescope on Kitt Peak in Arizona. The team sights, on average, about three near-Earth asteroids per month.

In one of the program's many successes, it discovered a number of new, erratically moving objects in the outer Solar System that have properties of both asteroids and comets. Belonging to what is called the Centaur family, they are made of both ice and rock, and measure many miles across. The Palomer survey discovered the prototype, Chiron. An impressive 150 miles wide, Chiron is larger than the states of Massachusetts, Rhode Island, and Connecticut put together. Because they are constantly jostled about by the larger planets, members of the Centaur group are expected to be prodded into an Earth-crossing orbit about once every thousand years. One shudders to think of an object like Chiron falling on our poor planet.

Naturally, a single program cannot spot every possible asteroidal or cometary threat to Earth. With limited resources, the Spacewatch team does the best they can to make sure that rogue intruders do not slip through their tracking net.

A new 1.8 meter (six-foot diameter) Spacewatch telescope is being prepared to come on line. With double the mirror diameter of the original (telescopes collect more light with larger mirrors), it will augment considerably the Arizona group's efforts to spot rocky wanderers in space.

In December 1994, team member Jim Scotti, employing the original (three-foot) Spacewatch Telescope, detected the closest asteroid ever discovered outside of the Earth's atmosphere. Passing within sixty thousand miles of the Earth directly over Russia, this object was found to be about thirty feet in diameter, roughly the size of a house. Fortunately, the body missed our planet by a safe margin. Otherwise, like the Wicked Witch of the East in the Wizard of Oz, someone might have looked up to discover a "house"—albeit a rocky one—falling down from the sky.

The Spacewatch 1.8-meter telescope is designed to detect asteroids that may threaten our planet. (Courtesy of Tom Gehrels.)

One of the most moving fictional descriptions of an asteroid impact on Earth was written by famed speculative novelist Arthur C. Clarke. In his epic work, *Rendezvous with Rama,* he ponders the devastation caused by the falling of a giant cosmic rock onto a highly populated region of Europe in the year 2077. He imagines such a shocking event would compel the human race to develop precautionary measures against future such disasters:

> Moving at fifty kilometers a second, a thousand tons of rock and metal impacted on the plains of northern Italy, destroying in a few flaming moments the labor of centuries . . . Six hundred thousand people died, and the total damage was more than a trillion dollars. But the loss to art, to history, to science—to the whole human race, for the rest of time—was beyond all computation . . .
>
> After the initial shock, mankind reacted with a determination and a unity that no earlier age could have shown . . . No meteorite large enough to cause catastrophe would ever again be allowed to breach the defenses of Earth. So began Project Spaceguard.[11]

Inspired by Clarke's bold prose, and motivated by the desire to avoid, at all costs, a real-life version of the disaster described in his novel, in 1993 a group of concerned astronomers from a dozen nations proposed the construction of a Spaceguard global warning network. In a series of meetings held in Erice, Sicily, and Tucson, Arizona, they outlined the details of such a system. In the scheme discussed, at least six or seven telescopes scattered throughout the world, equipped with state-of-the art electronic cameras, would be employed to track possible stray asteroids or comets.[12] Each component of the network would be in constant communication with the others. All data received would be corroborated and analyzed by sophisticated computer programs.

If one of the telescopes connected to the network detects a large asteroid or comet fated to pass dangerously close to Earth, then the entire system would be notified. The other elements of the network would continue to monitor the potential intruder, until the most accurate assessment of the danger could be made. If the news is especially bad, then governments around the world would be issued a dire warning and (one would hope) a plan for appropriate action.

Step-by-step, components of a global tracking system are falling into place. Programs such as the Minor Planet Center, working with groups such as Spacewatch, are performing a fine job pinning down the sizes and trajectories of thousands of potentially threatening objects. Eager to share their information with similar programs, they are laying the groundwork for a worldwide asteroid/comet information network.

As astronomers' capacity to track potential threats to Earth increases, the question arises of what to do if a body were found to be on collision course. Numerous ideas for destroying incoming objects have been advanced over the years. Renowned physicist Edward Teller once proposed that a well-placed nuclear bomb could blow an astral intruder to bits. As an alternative, Teller has also advocated building a laser defense system—an adoption of the "Star Wars" strategic defense program, perhaps—to pulverize dangerous extraterrestial bodies. Satellites would be placed into orbit to spot and destroy objects before they had a chance to harm us. Such systems designed to fragment encroaching bodies would be extremely expensive, however, costing many billions of dollars. They would also be dangerous, since pieces of the fragmented body still might plunge to Earth, possibly carrying with them, in the nuclear case, radioactive material.

Most astronomers currently advocate *diverting* encroaching comets or asteroids, rather than *destroying* them because of the expense and dangers associated, especially if nuclear missiles are employed. Assuming that there was enough time to take action, a well placed poke to the ribs of such an intruder, pushing it off its deadly trajectory, would be safer, cheaper, and more effective. The key to this strategy is to concentrate government resources in developing the best early warning system possible, including a complete database of all NEOs.

In the book, *Hazards Due to Comets and Asteroids,* several leading researchers, including Vadim Simonenko of the Russian Federal Nuclear Center, current director of the Space Shield Foundation, make a strong case for diverting threatening NEOs with what is called a "stand off" (slightly away from the object) nuclear or chemical explosion.[13] They argue the best way of pushing an intruder off its doomsday course would be to generate a 1 megaton explosion at a distance of a few feet away from its surface (though it may be hard to be so precise). The blast would heat up the object, expanding its surface, causing it to eject material. By physical principles, such as the law of conservation of momentum, the object would change its course enough to veer harmlessly off into space. The law of conservation of momentum is the same principle that causes a cannon to move backward once fired. In this case the cannon would be the asteroid or comet, and the cannonball, the hot material released after the explosion.

In another approach, the Air Force Space Command has advanced a detailed plan for diverting asteroids by hitting them with launched projectiles. It has assigned three of its associated laboratories the task of testing and perfecting such a program: Philips Laboratory in Albuquerque, Lawrence Livermore National Laboratory in California, and the Naval Research Laboratory in Washington, D.C. Together these centers have developed a $120 million test mission, called Clementine 2, aimed at determining how hard it would be to deflect an asteroid.

Stewart Nozette, director of the program at Philips, is optimistic that an asteroid on a collision course could safely be diverted. He has estimated a twenty to thirty ton projectile targeted at an asteroid three hundred to four hundred feet across could steer it away from Earth and propel it along a new, harmless path. The Air Force "wants to demonstrate that you can hit the things," says Nozette.[14]

To perfect an antiasteroid defense system, the Pentagon has planned to target three near-Earth objects with a barrage of ten pound guided missiles, each about three-feet long and one-half-foot wide. These projectiles would be launched from a spaceship, which would then move well out of the way and videotape what happens to the asteroids. The asteroids' amounts of deflection would be carefully monitored, and any material released during impact would be analyzed. Team scientists hope this data would help bolster their chances of creating a shield to protect the Earth from potentially harmful impacts—maybe even averting doomsday.

In late 1997, the United States government dealt the Clementine 2 mission a major, perhaps fatal, setback. President Clinton vetoed a $30 million congressional appropriation for the program. Reportedly, the president was afraid that funding the mission could possibly be construed as violating international arms-control accords.[15] The launch date of the Clementine 2 spacecraft, scheduled for mid-1998, has been indefinitely postponed. It is presently unclear whether or not the Pentagon will now be able to continue, in some form, its testing of asteroid deflection systems.

Prelude to Oblivion

If fate is unkind, one day you might wake up, put on your slippers, pick up your morning newspaper, and be astonished to read the headline: "Astronomers Detect Large Comet Headed toward Earth." After you rub your eyes a few times to make sure you are seeing correctly, and drink a fortifying hot beverage, you might then experience the horror of poring over an account of terrestrial disaster predicted to occur within a few months. "I knew they should have funded that defense system," you might mutter to yourself while shaking your head in disbelief.

Weeks later, a faint new body becomes visible in the sky—at first through binoculars, and then through the naked eye. As the pages of the calendar turn, it becomes brighter and brighter, until its ghastly figure is unmistakable—an orb of doom that outshines most stars. The comet has drawn so close you can almost sense the sting of its icy breath.

Doomsday now appears imminent. The streets are packed with drunken revelers, tossing down pints of strong smelling spirits as if there was no tomorrow. They jostle each other, occasionally brawling, uncaring about injury, convinced that only a fool would try to preserve his health and sanity in such a moment. Yearning for a sweeter oblivion than the one promised by the growing blaze across the sky, they clutch tightly onto the cheap comfort of their inebriation.

As you scurry through the town center trying to avoid encountering threatening drunks, your ears are assaulted by the repetitive, droning chants of a strange, new religious order: The Church of the

Blessed Survivors. Gaunt men and women, dressed in shimmering white robes, with images of comets shaved onto their nearly bald pates, form an otherworldly sight as they march down the crowded streets. They shout the cryptic slogan, "embrace the transformation," over and over again, until you want to scream.

A new sound does arrive: a thunderous boom straight out of Hades. A rumble in your heart, a churning in your stomach, and a dreadful feeling in your soul signal the time of doom is at hand. Thousands of tons of rock and ice have hit the Atlantic Ocean floor, setting off audible tremors around the world. From India to Indiana, from Togo to Tonga, from Austria to Australia, no one might avoid the somber tolling of armageddon's sonorous bell.

Reports of a gigantic tidal wave that has washed over the Statue of Liberty at almost three times its height crackle over the radio. The great tsunami has already flooded out most of Florida, Georgia, the Carolinas, Delaware, and New Jersey, and is well on its way to covering coastal New England, the Gulf States, and other places on the East Coast. The Atlantic Ocean has become bloated and fat, full of the buoyed artifacts of generations of family life.

The devastation is entirely democratic—neither the rich nor the poor are spared. Museum items, library books, scribbled grocery lists, and raw sewage float past each other, indifferent to their varying grandeur and places of origin. A recent Warhol exhibit from a museum of modern art ironically finds itself surrounded by a potpourri of rusty soup cans.

The skies have now darkened to a hazy twilight. Dust and debris unearthed by the cometary collision choke the atmosphere until it resembles the contents of an overused vacuum cleaner bag. Chilly winds and a light July snow signal the advent of a brutal winter without hope of spring.

Let us hope, if the great cosmic dice someday produce the unluckiest of rolls, and earthly civilization becomes imperiled with a fatal blow from space, a powerful enough defense system will be in place to insure that doomsday is avoided. One expects that if such a threat looms, a comet/asteroid diversion mechanism would quickly be put into place. Then, all of humanity might celebrate the vanquishing of the intruder and the renewal of life on Earth.

Or if all else fails, let us hope space travel will have advanced enough by then that our race might flee to safe haven on other planets. Although eventually— even if no comets or asteroids issue fatal blows—the world will end with the death of the Sun, space science should work to ensure that the demise of our home planet will not represent the death knell for our species.

CHAPTER EIGHT

ANGRY RED SUN

*. . . the sun, red and very large, halted motionless upon the
horizon, a vast dome glowing with a dull heat, and now
and then suffering a momentary extinction . . . [it] grew
larger and duller in the westward sky, and the life of the old
earth ebbed away. At last, more than thirty million years
hence, the huge red-hot dome of the sun had come to
obscure nearly a tenth part of the darkling heavens.*

—H. G. WELLS, *"The Time·Machine"*

Rendezvous with Cosmic Destiny

The ecology of our planet has been remarkably durable for
eons. While individual species have come and gone, Earth's
habitat has maintained itself in a livable state, more or less, for
millions of years. True, some geological eras have been better suited
for life than others. Environmental conditions in much of the world
are certainly more comfortable today than during the Ice Ages. Yet,
during harsh times as well as more pleasant times, the rudimentary
prerequisites for life have persisted.

Earth's ecological system has been so resistant to major changes
that only in recent decades has humankind acquired the ability to
alter it significantly and permanently. Sadly we now have the power
through global warfare and pollution to devastate the environment

and perhaps even destroy the world. However, if we avoid these pitfalls—and somehow manage to avoid (by science or mere luck) collision with stray astronomical objects, such as comets or asteroids—we might well expect Earth's ecology to maintain itself for many millennia to come.

We should not, though, by any means, be presumptuous enough to consider ourselves absolute masters of our own fate. Though we may work hard to save the environment, eventually life on Earth will be threatened by events beyond our control. Billions of years from now, the death of the Sun will transform our vibrant world into a lifeless chunk of frozen material and rock.

Earth's environment is like a tropical garden in Alaska, surrounded by a climate-controlled dome. As long as it receives a continual dose of heat and light energy, it can thrive. But if the power source (the Sun) that protects it from its harsh surroundings (space) were ever to become extinguished, its temperature would soon drop to a few degrees Kelvin (close to absolute zero). In short order, all life within its shielded domain would wither and die.

The Sun provides for Earth's vitality by infusing it with usable energy. This power is utilized by living organisms to create order from chaos by means of processes such as photosynthesis. Without sunlight, these constructive mechanisms would soon grind to a halt.

Life on Earth requires continuous injections of orderly energy to inoculate it against the creeping infestation called *entropy increase*. *Entropy* is the term used by physical scientists to refer to how disorderly a system is. A complex structure, such as a crystal or a DNA molecule, is said to have low entropy. In contrast, a random system, such as a diffuse mixture of gases, is considered to have high entropy. The law of increasing entropy, discovered by Rudolf Clausius in 1865, states that physical processes, left to themselves, tend to develop over time into states of greater and greater entropy. That is, without outside sources of order, all things must decay.

The human body represents a perfect example of this principle. When people are alive, they sustain themselves through eating and drinking. The food and drink they ingest is metabolized by their cells and generates the power needed for a wide range of bodily functions—bone healing, for example. Thus foodstuffs provide the order needed for life-renewing processes.

After death, the physical composition of human bodies rapidly breaks down. Once the body's system for utilizing orderly energy is no longer viable, the process of decay soon kicks in. Natural processes that consume a body—bacterial and fungal agents, for example—very quickly begin to take their toll upon the human form. Within days, few would recognize the deceased. Such is the shroud of anonymity woven by the cruel effects of increasing entropy.

The orderly energy provided by food is far from inexhaustible. Human sources of nourishment (plants and animals) are themselves dependent for their sustenance on other sources of usable energy, namely simpler forms of life. Ultimately—passing down the food chain—all animal life derives its nutrition through vegetation. Plant life, in turn, derives its power from solar radiation—through photosynthesis or another comparable process. In short, all beings on Earth obtain their orderly energy from the Sun, either directly or indirectly. And the Sun won't be around forever. Once it dies, Earth's food chain, and all life on our planet, will perish with it.

Shining down on us with its radiant power, the Sun might appear eternal. Yet the mechanism that enables it to give off seemingly endless quantities of light and heat is subject to the same law of increasing entropy that acts on Earth. Eventually, it, too, will run out of usable fuel and lose its effectiveness. The Sun will decline and pass away, causing life on Earth to cease.

To understand why the Sun's resources are finite, we must discern how it derives its energy. Science has shown it obtains its fuel from the process of nuclear fusion, the same dynamo that powers hydrogen bombs. Let us examine how the Sun fuses hydrogen to produce its radiation.

Inner Workings of the Sun

Until the twentieth century, the Sun's primary source of energy was unknown. When scientists first tried to address this issue, they speculated the Sun radiated because it was literally on fire, like a burning lump of coal or piece of wood. As in the case of coal or wood, they thought it obtained its power through a chemical process, such as oxidation. They sought in vain for a chemical reaction that would produce the mammoth quantities of heat and light radiated by the Sun. Gradually they realized that no such chemical mechanism exists.

In the mid-nineteenth century, Lord Kelvin (William Thomson) proposed, as an alternative mechanism, that the Sun releases radiation by slow contraction. His suggestion made use of the physical law that gases, when compressed, tend to heat up. This is the principle by which a pressure cooker works; as external pressure is applied, internal temperature rises.

In similar manner, Lord Kelvin postulated the Sun is slowly shrinking, converting gravitational energy to heat in the process. The Sun's giant ball of gas, he argued, acts as a colossal pressure cooker, causing its temperature to increase with time. The Sun's pressure, and consequently its temperature, continue to rise as it contracts further and further. As it grows hotter and hotter, it expels its excess energy into space, which we observe as sunshine.

Lord Kelvin estimated how long it would take for the Sun to release all of its gravitational energy in this manner. Based on his calculations, if the Sun produced its power exclusively by shrinking it would last twenty million years. Because he thought the Earth and Solar System were only several million years old, he considered this figure to be quite sensible. It seemed especially reasonable when compared to the biblical estimate that the world was created only several thousand years ago—a popular belief of his time that he tried hard to refute. Even if the Sun has lasted five million years already, he reasoned, it could conceivably shine by this method for many millions of years longer.

Today we know that Earth, the Sun, and the rest of the Solar System are almost five *billion* years old. The Earth's age has been gauged through the radioactive dating of rocks. By measuring, in ore samples, the ratio between certain radioactive substances (such as uranium) and their byproducts, and by knowing these elements' rates of decay, geologists can deduce the age of the samples. From this figure, they can place a lower limit on the age of our planet: approximately 4.6 billion years. Because current theories of the Solar System suggest all of its components were formed as part of the same process, astronomers believe the Sun's age is comparable to Earth's.

Lord Kelvin's shrinking Sun hypothesis could not account for such a long solar lifetime. Like a fully squeezed orange, the Sun would have long ago run out of juice. Its gravitational energy would have been pressed out of it billions of years ago. Therefore it would not have been shining when humankind developed and would not be shining today. Clearly, another explanation for sunshine is necessary.

In 1905, Albert Einstein's special theory of relativity was proposed, which provided the theoretical foundation for the modern science of nuclear fusion. His bold idea that matter and energy are interchangeable, embodied in his famous equation of energy–mass equivalence, pointed to ways in which each form could be freely converted to the other. Fusion, a means of deriving energy from mass by smashing atoms together, comprises a veritable alchemist's stone for generating tremendous quantities of force.

We have discussed how fusion can be used to design and build terrible weapons of mass destruction, possessing the power to envelope the Earth with a shroud of lethal radioactivity. The peaceful use of fusion as a fuel source is a scientific dream waiting to be fulfilled. Experimental fusion reactors have been built and run for decades, but none has proved efficient enough to be economically viable as a power station. Still, many nuclear physicists hold out hope that much of the world's energy needs will be fulfilled through hydrogen-powered fusion dynamos.

The energy source humankind has yet to perfect for its own use is nature's favorite powerhouse in the cosmos. Soon after Einstein's theory was developed, astronomers realized nuclear fusion is the way the myriad stars in the universe generate their mighty beacons of light. The conversion of stars' mass into energy by the combining and recombining of their atomic constituents enables them to pour their excess heat into space for eons and eons in the form of starlight.

The Sun, as a typical star, produces most of its energy through the common process of fusing hydrogen into helium. It possesses ample hydrogen fuel, enough to last it for many billions of years. Through this mechanism, it has survived far, far longer than it would have by simply exploiting the gravitational energy released by contraction.

Solar fusion takes place in a series of steps, each involving the release of excess radiation. Each stage triggers the next, in a multibillion-year-long chain reaction. Over time, like bricks stacked one by one to form the walls of a house, simpler elements within the Sun combine with each other to form increasingly complex substances. Unlike architectural construction, however, the solar building process manufactures energy, rather than consuming it.

The Sun's chain reaction starts when the nuclei (or cores) of two hydrogen atoms smash together, to form a particle called a deuteron. Deuterons constitute the nuclei of a substance called deuterium, more commonly known as heavy hydrogen. Almost twice as massive as ordinary hydrogen atoms, deuterium atoms consist of both electrically positive protons and electrically neutral neutrons.

In the manner so brilliantly predicted by Einstein, the mass difference between pairs of hydrogen nuclei and single deuterons is converted directly into radiation. In other words, because a deuteron weighs less than two hydrogen nuclei, the leftover mass turns into energy. This power is radiated outward into space in the form of sunshine.

By analogy, one might imagine a recipe for pound cake that calls for two eggs. Typically, if one were to add two fresh eggs to a cake batter, one would first crack them open, release their contents, and then discard their shells. Because the weight of the shells would not be counted, the egg mixture added to the batter would weigh less than the two original eggs. The mass difference—the shells—would now form part of the garbage. In similar fashion, the Sun's superfluous material—its "garbage"—is released as heat.

In the next stages of the solar cycle, similarly involving the liberation of excess energy, deuterons merge with hydrogen nuclei to form an isotope (type) of helium, called helium-3. Helium-3 nuclei combine, in turn, to form helium-4 (standard helium) particles. A helium-3 nucleus consists of two protons and one neutron; helium-4 contains an additional neutron.

Over time, more and more of the Sun's hydrogen fuel has become converted into helium. Helium itself, if subject to sufficient pressure, can fuse together into heavier elements, such as carbon and oxygen. These more complex processes occur within the Sun, particularly inside its core, but much less often than hydrogen fusion. During the solar lifetime, the dominant source of sunshine, by far, has been the multistep transformation of hydrogen to helium.

The Sun is hardly immortal. About five billion years from now, it will begin to run out of its primary fuel reserve. Switching from hydrogen fusion to helium fusion as its main source of power, it will start to blaze much brighter. Soon, it will pour out energy over a hundred times quicker than it does today. Issuing its banshee's wail of rapid, explosive burning, it will embark on its agonizing march toward its inevitable demise.

The Sun's journey toward death will be a prolonged one—lasting 100 million years from the time it has expended its original hydrogen. As it burns its fuel faster and faster, it will swell up and start to glow a fiery crimson. Ballooning outward until it has gobbled up the inner planets Mercury and Venus—and perhaps even the Earth—it will evolve into the type of stellar behemoth that astronomers call a *red giant* star.

Though red giants shine brightly, their glow, by stellar standards, is quite cool. They are so brilliant they can be seen with the naked eye thousands of light-years away. Yet, in spite of their substantial light output, they are considerably cooler than average—akin to high powered flashlights packed in ice chests.

The reason bloated stars burn in such a cool fashion is that their temperatures reduce when they expand. Like hot air balloons, they cool down as they get bigger—assuming that their internal fuel sources do not provide enough energy to compensate. By the time stellar bodies have reached red giant proportions, their temperatures have dropped thousands of degrees.

Gradually, the bloated red Sun will lose hold of its outer layers. A series of contractions within its central core—caused by the fact that its nuclear engines no longer will produce enough internal pressure to keep it stable—will generate strong vibrations. As the solar core collapses considerably under the influence of its own weight, powerful shockwaves will propagate outward from the Sun's center. These waves will push the loose gaseous material of the Sun's outer

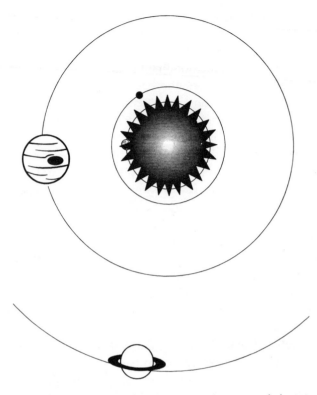

When the Sun swells up billions of years from now to become a red giant star, it will likely engulf many of the inner planets.

envelope off into space. Eventually, after millions of years, all that will remain of our mother star is its highly condensed core.

The Sun's remnant, at that point, will be tiny, dense, and dim: a celestial object called a *white dwarf* star. White dwarfs (such as the star Sirius B, discovered in 1862 by Alvan Graham Clark) are astronomical glowing embers. Like the relics of a newly extinguished campfire, white dwarfs glow in a faint, but scorching hot, manner. In this way they represent the antithesis of red giants; instead of being low temperature stars from which mammoth quantities of energy gush out, they are high temperature objects from which tiny amounts of energy trickle.

If all known stars could be lined up on a lot, white dwarfs would be the subcompact models. They are profoundly more condensed than typical stellar bodies. Earthlike in size, they have radii of just a few thousand miles. Yet they are so incredibly dense a single cubic inch of their material weighs hundreds of tons.

Astronomers predict the Sun will exist as a white dwarf for millions of years, gradually exhausting its last reserves of energy. Eventually, it will no longer have the power to shine and will become the charred relic known as a black dwarf. A cold, silent heap of rock, unable to announce its presence by emitting discernable radiation, the Sun's cosmic drama will finally draw to a close.

The Case of the Missing Neutrinos

How do astronomers know their models for stellar evolution and death are accurate? Obviously they cannot observe stars for billions of years, painstakingly checking to see if their predictions have proven correct. In fact, if the Sun somehow were to die right now, there would scarcely be a chance for scientific observers to cash in on their bets. Any written wagers on how long the Sun might last would be consumed by its lapping flames as it expanded to gigantic proportions.

Fortunately, there are concrete measurements researchers can perform to test theories of solar (and stellar) behavior. According to physical theory, atomic nuclei as they fuse together emit high speed particles called neutrinos. Neutrinos are extremely light objects, quite possibly massless. (Whether or not they have mass remains a matter of controversy. Recent experiments seem to indicate that they do have mass.) Veritable will-o'-the-wisps, they rarely interact with other particles, passing by most without even noticing.

Neutrinos are all around us. Second after second, myriads of them streak through space, plummet through the atmosphere, pass unhindered through each and every earth-dweller, zip through the Earth's interior, and eventually return to deep space. In the words of John Updike, "Earth is just a silly ball, to them, through which they simply pass."[1]

A significant portion of these invisible cosmic intruders can be traced back to solar fusion processes, having originated in the Sun's core and escaped through its surface. By measuring the rate of neutrinos that reach Earth from the Sun, particle physicists hope to test standard solar models.

The main difficulty researchers face in detecting neutrinos is that they interact so little with other forms of matter. Ongoing neutrino detection projects, such as Brookhaven National Laboratory physicist Ray Davis' experiment in the Homestake Gold Mine in Lead, South Dakota, employ gigantic vats of fluid to capture these elusive creatures.

The Homestake experiment has been operating for more than three decades. It is set deep within the ground, 850 feet below the Earth's surface, in order to screen out the unwanted interference of cosmic rays (high energy particles from space that continuously bombard Earth). Team researchers oversee a tank with over 100,000 gallons of cleaning fluid (perchloroethylene), looking for signs of neutrino interference. Each time a neutrino interacts with the liquid, a chlorine nucleus becomes induced to decay into radioactive argon—a process explained by nuclear theory. Detectors surrounding the tank monitor the fluid and record the number of such transitions. From this information, Davis and his team have estimated the rate by which solar neutrinos pass through Earth.

The Homestake group's figure, approximately two neutrinos every three days, is much lower than theoretical models of the Sun would indicate. Most solar models predict that 1.8 neutrinos per day would be trapped by the fluid. Since other teams have verified this discrepancy, observers are reasonably certain the Homestake results are valid. Therefore, astrophysical theorists are faced with the challenge of explaining the whereabouts of the missing neutrinos.

One explanation for the solar neutrino shortage is the temperature of the Sun's core is lower than previously estimated. Thus fusion processes there would run slower, and fewer neutrinos would be produced. Another favorite theory is that each neutrino has a tiny mass, and has a two-thirds chance of transforming into a particle that observers cannot detect with the gold mine apparatus. Consequently only one-third of the neutrinos originally produced through solar fusion are detected. To resolve the longstanding dilemma, researchers are currently testing these and other theories.

Once the mystery of the dearth of solar neutrinos is resolved, theorists will be able to develop more realistic models of the evolution of the Sun. Then we will have a deeper understanding of solar processes and greater insight into the events that will portend the demise of our parent star.

Ultimately scientists will be able to predict with great accuracy when the Sun will die. One hopes this will allow the human race ample opportunity to flee a dying solar system and set sail toward healthier worlds, assuming the technology to do so becomes available. The more we learn about the Sun, the greater the chances that doomsday—due to its death—might be postponed.

Celestial Timebombs

As a star of average weight and composition, the Sun will likely follow a largely uneventful, quiescent path toward death. Stellar bodies of its mass and type generally share its calm fate of quietly evolving into white dwarfs and then slowly fading into extinction. Chances are, no celestial fireworks will mark the Sun's ultimate death throes.

If the Sun today were a few times more massive, however, astronomical research indicates its final stages of existence would be far more violent. Stars several times heavier than the Sun eventually undergo fantastically powerful supernova explosions. In these sudden blasts they give off, for a brief interval, more radiation than the total output at that time of the rest of the galaxy's stars.

Before a star explodes in this manner, it must pass through a series of evolutionary stages—similar to, but more intense than, those which the Sun will undergo. Like the Sun, a heavy star will spend its youth generating massive quantities of energy by fusing hydrogen gas into helium. It will shine that way for hundreds of millions of years, until it has mainly used up its primary fuel reserve. When that happens, the cooler, loosely held material in the outer regions of the body will dilate, and a large gaseous envelope will form. The wispy layer will grow more and more immense, until the star has expanded far beyond its original size. The resulting colossal object will be what astronomers call a *supergiant*: a star that, in terms of size, is to a red giant what a plump tomato is to a meager cherry.

Because supergiants are so much bigger and heavier than ordinary red giants, their cores are much denser and therefore substantially higher in temperature. These hot centers act as great cosmic melting pots to blend together simple atoms into more complex structures, releasing energy in the process. As hydrogen, helium, and other light elements drip into these boiling cauldrons, they fuse together into heavier materials, such as carbon, nitrogen, oxygen, silicon, and sulfur.

Until the core material is exhausted, this fusion process automatically repeats itself. Each time a merger of two atoms occurs, radiation is released. This radiation, in turn, heats up neighboring particles, increasing the chances they will combine as well. Over time, the stellar core is churned over and over, wresting more and more energy from its hot, compressed fuel.

Ultimately, this chain reaction must cease. The most massive element that can be forged by the stellar fusion process without the addition of extra energy is iron. Once iron is created, the supergiant's core can no longer power itself by fusing together atoms. Iron and other heavy elements won't combine on their own; additional energy needs to be supplied.

Once further fusion becomes impossible, the supergiant's central region suddenly becomes unstable. Since no more power can be generated there by atom smashing, the only way for the core to produce additional energy is to collapse. Like a balloon with its air quickly let out, that's exactly what it does. In a flash, it shrinks down to a dense ball less than seventy miles in diameter.

The sudden collapse of a supergiant's core results in a phenomenal release of energy. Enough power is generated to blast the outer region of the star millions of miles across space. Such a supernova explosion can be astronomically observed as an extraordinary increase in light intensity from the part of the sky where the supergiant used to reside.

We need not worry about the Sun suddenly exploding. Its low mass makes it an unlikely candidate for a fiery supernova blast. Unfortunately, the Sun's relative stability does not make us immune to the effects of such explosions. Nearby supergiants, poised on the brink of blasting themselves to smithereens, present unmistakable perils to our world. Each is a ticking timebomb with a frightful potential for terrestrial destruction.

If a supernova explosion were to occur less than fifty light years (three hundred trillion miles) from Earth, scientists believe complex life on our planet would be decimated. Intense radiation would saturate Earth's atmosphere, precipitating terrestrial conditions akin to those inside a microwave oven. Any earthly creature for which concentrated radiation is poison—and that includes most beings would not last long when subject to such a horrific barrage from afar. Because of radiation-induced mutations, the offspring of those life forms that did survive would stand a high chance of being born with extra limbs or with impaired organs.

Imagine the grave state of our planet after bombardment by high energy radiation from a nearby stellar explosion. The bulk of the human population—perhaps everyone—would be killed by burns, radiation poisoning, and skin cancer. Little surface vegetation or animal life would be left; anyone who managed to survive the blast would likely die from starvation. However, because of their astonishing ability to withstand harsh conditions, insects would likely thrive—though possibly in mutated versions. Seven-legged ants, ghostly white cockroaches, wingless crickets, and other strange crawly creatures would join their more conventional brethren as undisputed masters of Earth's surface. Life deep beneath the ocean waves would probably also continue to do well. This, of course, would present little consolation to the ravaged human population.

Escape from such an ordeal—assuming a nearby supernova explosion could somehow be predicted—would require abundant planning and heaps of good luck. If our planet were fated to be bombarded by such a fire ball, the human race would need substantial warning if it were to have any chance to flee Earth. Today, astronomers would be unable to make such a firm prediction; perhaps they would be able to in the future. A supernova's energy pulse, traveling at the speed of light, could not be outrun. Unless we had a significant head start, in a speedy course toward a distant new planetary system, we would surely be destined to perish.

None of the stars within fifty light-years of the Solar System seems massive enough to blast apart in the manner we have discussed. Such an explosion, involving the sudden collapse of a supergiant, is called a Type II supernova. However, a second variety of supernova, called Type I, does appear possible in our vicinity.

A Type I supernova occurs when a white dwarf and a red giant form a binary pair. Gradually, the white dwarf accretes (pulls in) more and more of its companion's material, becoming more and more bloated in the process. Like a spoiled child, it keeps wanting more and more to eat—even as it gets heavier and heavier. Eventually, like a character from a Roald Dahl children's story, the white dwarf's greed leads to its catastrophic undoing. It utterly and completely explodes, leaving not a scrap behind. The amount of energy it releases is enormous—about 10^{30} (a million trillion trillion) times greater than that of a typical hydrogen bomb.

White dwarfs are so faint astronomers are uncertain exactly how many lie in our neighborhood and, of these, how many could potentially explode. Therefore, they cannot predict when a devastating blast from a nearby supernova will treat our planet like a microwave meal. Let us hope for the best, and offer a toast to the health and long life of our stellar neighbors.

History tells us that many supernova, seen as fiery stars, have been observed. Fortunately, none have been close enough to affect our planet. Nevertheless, their visual impact upon the unprepared skywatcher—not expecting to see such a bright object in the heavens—must have proven quite alarming.

In 1054, Chinese observers noted a brilliant flash of light in the constellation of Taurus the Bull, where none had been seen before. That sighting was the one of first recorded observations of a supernova explosion. The apparition they called a "guest star" lingered in the sky for fully a year before it faded completely from view. At night, it lit up the sky as if it was a second full moon. It was even visible in broad daylight for almost a month. The strange nature of its unanticipated appearance and disappearance suggested it was a celestial oracle of doom, rather than simply a natural occurrence.

Surprisingly, European chronicles from that period do not mention such a cataclysmic event. Aside from the Chinese description and an account by Moslems observers, the only other depictions of the supernova of 1054 can be found in the imagery of several native

American rock drawings. With radioactive carbon-14 testing, geologists have determined these paintings date back to that era. The artists who drew these renditions were Anasazi, ancestors of the Hopi tribe who lived in the region that is now Arizona and New Mexico. On cliff and cave walls they painted scenes of a bright new star next to a crescent moon. In these pictographs, the relative positions of the star and moon are exactly what astronomers would expect.

Today we know that the 1054 explosion constituted the sudden death of a massive supergiant approximately five thousand light years (thirty quadrillion miles) away from Earth. The event left two unmistakable souvenirs: one, called a *supernova remnant*, that can be visually observed with a telescope, and another, called a *pulsar*, that makes itself known through pulsing radio signals. The former object consists of the gas blown off in the blast. These emissions settled into

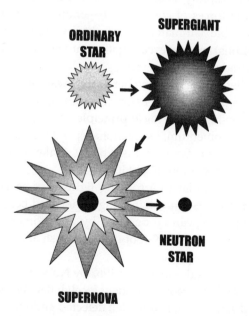

ORDINARY STAR

SUPERGIANT

SUPERNOVA

NEUTRON STAR

Depicted here are the stages in the formation of a neutron star. After a massive star swells up into a supergiant and then explodes, its collapsed core (shown as a black dot) becomes a rapidly spinning neutron star, also known as a pulsar.

a hazy, multicolored formation known, because of its crablike shape, as the Crab Nebula. In addition to the Crab Nebula, many other wispy supernova remnants can be seen in the sky, colorful relics of similar explosions. The second byproduct of the supernova of 1054 is far less obvious. Rather than comprising a flamboyant splash of colors, it is an emblem that is far more demure. The Crab pulsar, the tiny, rapidly-spinning object that constitutes the relic of the dead supergiant's core, can only be observed by scientists monitoring its regular emissions of radio signals.

Pulsars are created in the process of catastrophic stellar collapse, when the star's central section runs out of fuel and suddenly draws itself inward. Like a weightlifter about to raise a heavy barbell, in a brief instant it sucks in its massive iron stomach. As the star's belly contracts tighter and tighter, the implosion generates enormous internal pressure. Rapidly its innards become pulverized by the astronomically large forces of compression—enough to crush a freight train into a thimble. Ultimately the core's constituent particles (protons and electrons) are squeezed together into neutrons. The core has become an ultradense neutron star.

Like a ballet dancer drawing her arms inward while pirouetting more and more rapidly, the neutron star spins faster and faster as it contracts. Following the physical principle known as the conservation of angular momentum, its rotational speed increases in proportion to its diminution. Soon (if it is typical) it takes only seconds or less to complete revolutions about its axis.

In a manner similar to a rotating lighthouse beacon, the collapsed star broadcasts a revolving, highly focused beam of energy. This pulsating signal can be detected by astronomers' radio telescopes, such as the giant 1000-foot-diameter dish at Arecibo, Puerto Rico—hence the name pulsar.

Pulsars were first discovered in 1967 by Nobel prize winner Anthony Hewish and his graduate student Jocelyn Bell. While observing a region of a distant constellation, Bell detected periodic variations in its radio output. She was astonished to find such a regular signal in space. After ruling out mundane sources, such as nearby radio transmitters, Bell deduced the broadcast was extraterrestrial in origin. After bringing the discovery to Hewish's attention, they speculated, half in jest, that an alien civilization was sending out messages. Finally they realized that a rotating neutron star formed the signals' origin.

If the core of an especially massive supergiant collapses, then an object even most compact than a neutron star is formed, called a black hole. Unlike the case of neutron stars, where contraction stops once all of the protons and electrons in the core combine to form neutrons, black holes contract even further. The core neutrons themselves smash together, forming a tight ball of matter so thick it constitutes the densest known body in the universe.

Gravitational theory mandates that massive objects generate powerful forces of attraction. Consequently, the attractive properties of black holes are tremendous. Like celestial Venus-fly-traps capturing ill-fated insects, they trap everything that ventures near. Not even incoming light can escape their grasp. For this reason they are invisible, or "black," hence their name.

Depending upon the mass of the original star, each time a supernova blast occurs, either a neutron star or black hole is produced. The 1054 explosion, for example, produced the Crab pulsar. A much more recent example that was observed in 1987, Supernova 1987a, is similarly believed to have created a rapidly rotating compact remnant.

The 1987 supernova formed the perfect emblem of stellar mortality. Suddenly and unexpectedly, a giant star 100,000 light years away in a region of the night sky near the Tarantula nebula underwent catastrophic collapse. Because the explosion took place closer to our galaxy, the Milky Way, than any other supernova since 1604, it became the most studied such event in history.

The blast was first observed by technician Oscar Duhalde, of the Las Campanas Observatory, who noticed while looking up at the stars during a coffee break that the Tarantula Nebula seemed unusually bright. He did not report his observation, however, until he found out Canadian astronomer Ian Shelton had photographed the same event. Shelton had noted a bright flash on an image he had taken of the Large Magellanic Cloud using the small, ten-inch lens telescope at Las Campanas. Meanwhile, around the same time on the other side of the world, amateur astronomer Albert Jones had ob-

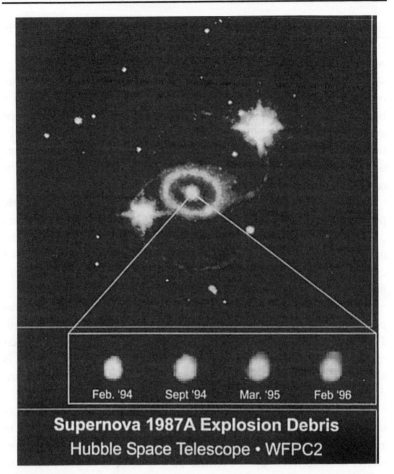

This photograph taken by the Hubble Space Telescope shows Supernova 1987a and its surroundings. The four panels show the evolution of the blast and its debris from February 1994 to February 1996. (Courtesy of NASA.)

served the same event in New Zealand. Because of the near-coincidence of these three astronomical observations, credit for the discovery of Supernova 1987a was shared more-or-less equally.

It is sobering to imagine the catastrophic impact of such an explosion. If there were planets circling anywhere near the star, they were surely demolished in the blast. Moreover, any planetary system within several trillion miles of the explosion would be doomed. Rushing outward at an enormous speed, radiation from the supernova would decimate worlds around proximate suns. One can only hope no stars explode in our own vicinity, for that would certainly mean the demise of Earth as we know it.

Requiem for a Star

A popular song from the 1970s states that burning out is better than fading away. This maxim may sometimes be true for music stars, who often become legends if they expire at the peak of their popularity. Considering, though, our precarious position on a planet orbiting another kind of star—the Sun—let us be thankful that astronomers expect it will fade away very slowly, not suddenly explode. Indeed, the Sun is expected to remain pretty much like it is for billions of years, until it finally swells up into a red giant.

Though the Sun's death will be slow, it will nevertheless represent the end of life on Earth. This demise will occur in one of two ways: either the Earth will be consumed in a pillar of fire as the Sun swells to gargantuan proportions, or it will be frozen over when the Sun's core recedes to form a cool white dwarf. Neither scenario would be pleasant for any life forms, including humans, that have survived on our planet until that fatal time. The Earth's oceans and atmosphere—its lifeblood—will either boil away, drawn up into space by intense heat, in the former case, or fall as solid ice to the ground in the latter. The rest of our planet, its mineral portion, may survive as a rocky sphere, bereft of all forms of life, from plankton to primates. However it could hardly be called the same Earth.

Perhaps no one conveys the sense of desolation at the end of a sun-shattered world better than H. G. Wells in his epic conclusion to *The Time Machine:*

> The darkness grew apace; a cold wind began to blow in freshening gusts from the east, and the showering white flakes in the air increased in number. From the edge of the sea came a ripple and whisper. Beyond these lifeless sounds the world was silent. Silent? It would be hard to convey the

stillness of it. All the sounds of man, the bleating of sheep, the cries of birds, the hum of insects, the stir that makes the background of our lives—all that was over. As the darkness thickened, the eddying flakes grew more abundant; and the cold of the air more intense. At last, one by one, swiftly, one after the other, the white peaks of the distant hills vanished into blackness. The breeze rose to a moaning wind. I saw the black central shadow of the eclipse sweeping toward me. In another moment the pale stars alone were visible. All else was rayless obscurity. The sky was absolutely black . . . Then like a red hot bow in the sky appeared the edge of the sun.[2]

Someday, scientists will likely develop interstellar travel and thereby liberate humankind from its earthly confines. Then, the thought of the Sun dying will no longer be as terrifying. As soon as our parent star threatens to lap up Earth with its flames, or otherwise render our world uninhabitable, we would simply exercise the option of relocating our civilization to another planet, or set of planets, orbiting new suns many trillions of miles away. Like nomads moving on to another oasis when their haven from the desert heat dries up, we would search the heavens for a new "watering hole," and then attempt the long journey there. In such a manner, our race could conceivably survive the transformation of its home planet's primary power supply into a bloated red giant, and then a dim white dwarf.

Fight, Flight, or Fright

We are voyagers at heart, a race of eternal wanderers. From early times, we have journeyed from place to place in search of shelter, sustenance, and freedom from danger. Throughout the eons, human circumstances have changed, but our zeal for new horizons has remained constant.

In an option that dates back to the dawn of history, if things seem to be going badly, people seek to relocate to virgin soil. Early hunters and gatherers would move on to new locales whenever they found themselves overwhelmed by adverse conditions. If torrential downpours turned an embankment into a flood plain, then they would seek higher ground. If drought transformed a lush meadow into a virtual desert, then they would set out to find new arable enclaves. If novel predators arrived on the scene, and tribal members considered themselves outnumbered, then the lands on the other side of the mountains might come to seem especially attractive.

With the advent of fixed-plot agriculture, mobility grew much more limited. It was harder to abandon a meticulously cultivated field than to leave behind a fruitful part of the forest gone sour. Yet, as Bible stories and other ancient chronicles report, it was not out of the question for those wedded to the land to become nomads if their fertile grounds grew fallow, and to resume farming— perhaps somewhere else—once circumstances grew more auspicious.

When the Old World grew crowded and depleted, European settlers ventured to the New World. They encountered peoples there who, long before, had ventured across a land bridge from Asia. America, with its pristine uncharted territories, was an irresistible draw to those wishing to escape from more battered lands.

Flight is not the only possible reaction adversity might engender. Depending on the situation, another valid option is to fight. As humankind's capacity for toolmaking grew more advanced, natural threats could be increasingly countered by use of proper ingenuity. For example, rather than abandoning their low-lying farmlands to the ravages of the sea, the Dutch constructed dams for protection against floods. Instead of shunning disease victims—as did medieval nobility to those afflicted with the plague—or sending them off to colonies, the healthy now help the ill battle their ailments with modern medicine.

Lamentably a third possibility, besides fight or flight, has existed since time immemorial. Psychologically, if a danger seems too severe, and defense or escape seem unrealistic, many become paralyzed with fright. Hence, the historical propensity for the rise of apocalyptic movements during eras of great adversity.

Consider what might happen if the Earth was challenged with unprecedented cosmic catastrophe. As in past crises, such as the bubonic plague in medieval times and the nuclear test of nerves during the Cold War, doomsday sects would no doubt arise and attract bewildered supporters. Fanning the flames of panic, they would likely counsel against too much reliance on science. Yet, at that pivotal moment, science might provide relief for the world's despair. Like the builders of dams or the discoverers of remedies, scientists might provide the means to fight against whatever ails our planet. For example, a defense system might be developed to deflect an asteroid before it collides with us.

For some types of astral disasters, such as a nearby supernova explosion, there would not be much science could do to mitigate the effects. If astronomers' predictive models (that they may have developed by then) anticipate such a catastrophe in time, flight from Earth, if possible, would be the only reasonable option.

Assuming that space science enables us to do so, when Earth is no longer habitable, we would have long set sail for new planetary systems. Vast space arks, perhaps, with room for many thousands of emigrant explorers, would insure that the human race continues its saga on new worlds. The enormous ships that convey our descendants to safety would constitute miniature Earths, harboring representative plants and animals brought along for sustenance and companionship. Ultimately, they would also serve to regenerate their kind on the planets upon which our species come to settle.

Thousands and thousands of generations might live and die on the space arks without ever setting foot on terra firma. Children would be born who knew of the blue skies and green grass of Earth only through weathered tales and ancient recordings. They will be as well adapted to space as we are to land.

When these space arks reach habitable new worlds, there would be cause for celebration. Like the early American settlers, the human race would reestablish its roots on alien soil. If other intelligent species already live on these planets, let us hope they would welcome the refugees in peace.

As cosmic time rolls on, many of the new settlements would be faced with their own world-destroying cataclysms. Supernova explosions, or other sorts of stellar death throes, might equally well demolish those systems as they would have ours. Then, once again, it would be time for our progeny to resume nomadic status, and find salvation elsewhere.

Eventually, our descendants' wanderings would necessarily come to a halt. The law of increasing entropy provides a time limit to the universe. When the last sources of energy in space become unusable, there would be nowhere to run. The "heat death" of the universe—the state of maximum disorder predicted by physics—would represent the ultimate triumph of the grim reaper. All cosmic processes, including the marvelous mechanisms that we know as life, would eventually cease. During the timeless season of the end, only a vast cemetery of lifeless orbs, the relics of once-vital matter, would remain in the inert cosmos. No mourning chimes would follow the last tick of our clockwork universe.

CHAPTER NINE

WITH A BANG OR A WHIMPER

THE END OF SPACE

This is the way the world ends . . .
Not with a bang but a whimper.

—T. S. Eliot, *"The Hollow Men"*

Heat Death

Like an epic Russian novel, the universe is filled with myriads of players, plot twists, and possibilities. Even the great Tolstoy, after penning countless paragraphs of lively prose, eventually sought to bring his chronicles to a conclusion. That splendid masterwork, the cosmos, the font of all ideas and the realm of all activities, only holds a finite number of pages to its script. Over time, the hand that composes cosmic history will cease to introduce new characters and write in new situations, until the last celestial persona dies out and no more lines are written. Once the great book is finished, nothing more will need be said, and time itself as we know it will cease.

The law of increasing entropy, as formulated in the nineteenth century by Rudolf Clausius, states that all processes tend to generate more disorderly energy than they use. Distinguishing between two types of energy, work (energy employed in achieving a goal—to move a boulder, for instance) and heat (wasted energy that dissipates as exhaust), he found that no process can convert heat completely into work; some heat must be left over. Thus, there is no such thing as a perfectly efficient engine; each must produce some exhaust.

It would be impossible, for example, to drain the ocean of all of its energy by turning all of its heat into work. If such a conversion process were realizable, the world's power needs could be met by inducing seawater to grow colder and colder, and then siphoning off the excess energy. According to Clausius, no engine could run by exploiting the bulk of the ocean's thermal energy; the perfect device might only utilize a small part of it.

In another expression of Clausius' principle, over time, any closed system (exchanging no material with the outside world) must naturally proceed from orderly (lower entropy) to disorderly (higher entropy) states, rather than the converse. In any natural process, entropy cannot decrease; it can only grow, or, at best, stay the same.

The tendency for greater disorder provides an unmistakable arrow of time. A natural series of events viewed forward in time tends toward disorder; it would only seem to proceed toward greater order if seen in reverse chronology (when a film or video runs backward, for instance).

A ceramic flowerpot falling off a high windowsill would shatter into a jumble of pieces. Any dirt contained within it would likely smear all over the floor, creating a mess. Such a scene would be most natural, albeit untidy. We would be astonished, however, if the reverse occurred: loose soil from the floor along with broken shards, assembling itself into a filled pot, and then rising onto the windowsill. Such is the absurdity of a scene in which entropy decreases on its own.

The best possible engine, according to Clausius, takes advantage of a situation in which an orderly system, in proceeding toward disorder, converts some of its energy into work. An example of this is a waterwheel. As flowing water passes over a turbine, it passes from an orderly to disorderly state, turning the wheel in the process. However, once no more water is left to drive the turbine—if, say, the river dries up—that engine would cease rolling.

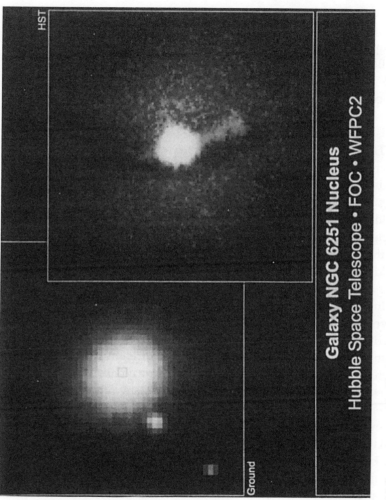

Galaxy NGC 6251 Nucleus

Hubble Space Telescope · FOC · WFPC2

Light pouring out of a black hole located in the NGC 6251 galaxy, three hundred million light years away in the constellation Virgo. (Left) The ground-base telescopic view of the galaxy's core. (Right) A composite image taken with Hubble's Wide Field Planetary Camera and its Faint Object Camera. The bright white spot at the image's center is radiation emitted by debris plunging into the galaxy's central

Stars are engines too, building up waste material as they function. As discussed, all stellar bodies, including the Sun, must eventually run down. Three known stellar end states are possible: white dwarfs, neutron stars (pulsars), and black holes. A star's ultimate fate depends on its mass. Only the heaviest suns, at least three times more massive than our own, follow the most mysterious path and end up shrouded forever from the rest of the universe.

Black holes are so dense—and therefore so gravitationally powerful—that not even light might evade their tight grip. As a consequence, the physics community used to believe black holes could never lose mass or energy, and must always remain intact. In other words, they were thought to be immortal.

The notion of black hole permanence seemed to violate the law of increasing entropy. If black holes did not decay, they could maintain their state of high order forever. Furthermore, if a disordered object, such as a shattered flowerpot, plunged into a black hole, it would be absorbed beyond recognition. Assuming the black hole did not gain entropy as a result, because the object's entropy would be lost, the total amount of entropy in the cosmos would diminish. This would violate the principle that the entropy of a closed system can never decrease.

In 1976, esteemed British physicist Stephen Hawking delivered a revolutionary talk, entitled "Black Holes are White Hot," and jabbed a white hot poker through previous notions about the subject. Making use of results developed by then-Princeton University research student Jacob Bekenstein, Hawking showed black holes possess entropy and temperature. Therefore, like any other bodies of nonzero temperature, they tend, over time, to radiate away their energy into space.

The measure of a black hole's disorder is the size of its surface area. Like the Blob, of B-movie fame, whenever a black hole swallows an object, it grows in girth. Its surface gets bigger in direct proportion to the entropy given up by the devoured body. Thus the entropy is not lost; it is converted into a different form.

If black holes consume everything, how might they emit energy? The answer to this seeming paradox lies in the field of quantum mechanics. Quantum theory says that there is no such thing as a perfect vacuum. Even so-called empty space is a sea of subatomic particles. These particles flit in and out of reality, like catfish jumping momentarily out of a stream and landing back in it.

For reasons explained by quantum theory, each time a fresh particle escapes from the void, it appears with its almost-twin: an antiparticle. Antiparticles have most of the same properties of their corresponding particles, except they possess opposite charge. Whenever an electron appears, for example, a positron (positively charged particle, otherwise identical to an electron) also arises.

Once a particle–antiparticle pair is created, it is usually annihilated almost immediately, as if it never existed. Like two evenly matched prize-fighters vying in a boxing ring, they can only stand so long before knocking each other out. However, if a black hole is around, one member of the pair—the antiparticle, say—might get devoured, while its companion survives.

Without its rival around to knock it out, the companion would escape unscathed. Leaving the black hole's vicinity, and venturing off into space, the surviving particle would appear to be radiating away from the black hole. Thus, by this process, black holes might radiate.

Black holes lose their energy extremely slowly. An average sized object would remain essentially intact for countless trillions of years, trickling out fuel bit-by-bit like an oil tank with a minuscule nick on its side. Gradually, however, after innumerable years have passed, nothing would be left of the once mighty body; it would have irradiated itself into nonexistence.

In short, the cosmos in its final state of "heat death" (maximum disorder) would resemble an automobile junk yard of immense proportions. The vacuum would be littered with parts and pieces of old clunkers: the corroded remnants of once vital bodies such as stars and planets. Even these relics would decay into nothingness, given enough time.

Might there be a way out? In Isaac Asimov's classic story, "The Last Question," characters ponder whether or not entropy might be reversed and heat death thus avoided. At first this question is posed hypothetically. Throughout the tale, as the universe winds down, the issue assumes increasingly desperate significance. Eventually, after all hope is lost, and the cosmos is almost at the point of extinction, a supercomputer finds a brilliant way of reversing entropy's course and reanimating the universe. "Let there be light," the computer bellows, and miraculously the cosmos is young again.

It need not take a miracle for the universe to be saved from an agonizingly slow heat death. Physics provides several possible escape clauses. First of all, the law of increasing entropy has been shown to apply to finite systems, but not necessarily to the cosmos as a whole (though it is generally assumed to do so).

Second, the laws of physics might change over time, with new principles taking effect in the latter stages of the universe. Humankind has witnessed so far such a minute portion of the history of the cosmos; therefore one cannot rule out the possibility physics could be radically different in the distant future. Conceivably, new physical laws might slow down or reverse cosmic decay.

Third, cosmology tells us that if the density of the universe is great enough, then long before heat death does it in, it will shrink down to a mathematical point—and possibly expand again, revitalized. This last possibility, an issue of vital interest to cosmologists, calls for full discussion.[1]

The Shape of Cosmic Fate

As Einstein pointed out in his remarkable theory of general relativity, published in 1916, space is not just a flat stage upon which the cosmic drama is acted out. Rather, it is a dynamic participant in its own right, responding to the exertions of astral actors with its own movements. Like the Globe theatre of Elizabethan times, it has its trap doors and moving platforms that contort in response to the cues of players.

Space, Einstein theorized, becomes warped in the presence of matter. This curvature takes place in a higher dimension than the usual three: length, width, and height. One might wonder how to picture such a novel dimension. Since we cannot fathom it in the manner we might envision a square or a cube, we are forced to rely on three-dimensional analogies.

Einstein found that the amount of curvature in a particular segment of space depends on its material content. The more mass in any given region, the greater that region curves. The Sun, for example, as a massive body, distorts the spatial domain in which the rest of the Solar System lies, causing it to assume a bowl-like shape. All of the planets, asteroids, and comets whirl around the basin like water swirling around a drain.

The Sun's dent in the fabric of the universe is not deep as those created by denser bodies. In the extreme case of a black hole, the depression carved out in space by its highly concentrated mass is so steep not even light might escape. Instead of a gently curved bowl, think of a long thin funnel, collecting all matter and energy in its vicinity. Such is the affect on its surroundings of an object of unimaginable compactness.

After he postulated the principle of general relativity, Einstein sought to apply it, not just to celestial bodies such as stars, but also to the cosmos as a whole. He expected to prove that the universe, on the largest scale, is flat and unchanging, like a fixed tapestry. Instead, he was astonished to discover his theory produced a radically different prediction: a wholly dynamic universe, forever expanding outward. The real universe doesn't get bigger or smaller, he thought, imagining that he had made a mistake with his dynamic model. To "fix" the solution he added an extra, stabilizing term (called the cosmological constant) to his equations governing the behavior of space. He announced this modified approach in 1917.

Russian physicist Alexander Friedmann was curious to explore the implications of Einstein's theory without the cosmological term. He wanted to find the broadest set of solutions that would satisfy Einstein's principles and still reproduce the way the contemporary universe appears. In particular, he noted the cosmos appears to be isotropic (all directions in space looking outward from any vantage point appear basically the same), as well as homogeneous (examining them on the largest scale, all regions of the cosmos have essentially the same distribution of material). Therefore, Friedmann looked only at isotropic, homogeneous cosmologies.

The concepts of isotropy and homogeneity are fairly simple to understand. Imagine standing at the intersection of eight busy highways. No matter which way you turn, you see the same thing: busy streams of cars moving in equal quantities either away from you or

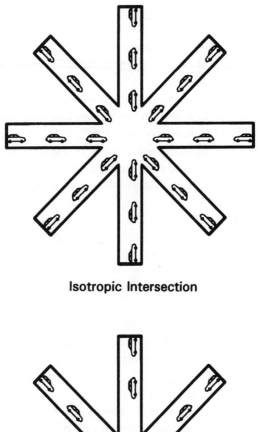

Isotropic Intersection

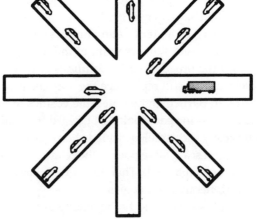

Anisotropic Intersection

Isotropic versus anisotropic intersections of highways.

toward you. Such a scene is isotropic. You then walk to another intersection, and you witness an identical picture to the first: eight busy highways radiating away. Eventually, you explore every single intersection in town, and realize they all appear exactly the same. Such a town is homogeneous.

In 1922, Friedmann published his results. He found three distinct cosmological solutions, referred to as closed, open, and flat models. In the closed model, space wraps in on itself in the manner of the higher dimensional equivalent of a great sphere. If one travels through such a universe for a long enough time, one eventually arrives back at one's starting point.

The open model, on the other hand, is saddle-like in appearance (the higher dimensional equivalent of a saddle, that is). Such a universe has no limits; one might venture an infinite distance in any direction without returning to one's point of origin. Finally, the flat cosmology is simply an endless smooth plane.

The average density of the universe governs its type. If the cosmic density (the amount of matter per unit volume in the universe) lies above a critical amount, the universe is closed. If it equals that amount, then it is flat. Densities less than the critical quantity mandate an open cosmos.

Friedmann further found the long term behavior of the cosmos depends on to which geometry it corresponds. Closed universes expand outward from a point, reach a maximal size, and then contract in again. Open cosmologies expand forever; they never reach a maximum volume. Finally, flat spaces teeter forever on the edge of either expanding forever or contracting down again. A minor mass fluctuation might eventually force a flat universe to collapse; otherwise, like an open cosmos, it would expand indefinitely.

Friedmann's models were considered purely hypothetical at first. Often, like customers at an electronics store pressing buttons and turning dials to check out new gadgets, theoretical physicists like to toy with the parameters of a new theory to see what might happen. In formulating his models, Friedmann was particularly interested in testing the implications of general relativity. Therefore, he applied it to several hypothetical circumstances.

In the late 1920s astronomer Edwin Hubble showed that the Friedmann models were not just theoretical; they genuinely seemed

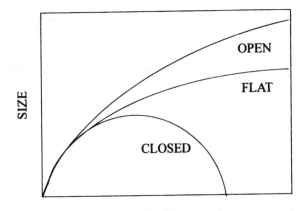

Friedman models of the universe.

to describe the dynamics of the cosmos. Hubble's discovery that the universe is expanding ranks as one of the most important scientific findings of all time. Since the universe, as a whole, is isotropic, homogeneous, and presently getting bigger all the time, Hubble's observations joined Friedmann's models and Einstein's original theory in paving the way for modern cosmology.

Hubble's discoveries radically altered the prevailing scientific views of space, time, and the cosmos. Before Hubble, the accepted picture of the universe was a static, eternal arrangement of stars. Few in science imagined that time and space might have natural limits. Hubble changed all that by revealing the cosmic panorama was not just immense, but expanding as well.

In 1924, Hubble made his first major breakthrough, proving that the universe's stellar constituents are clustered into vast distant galaxies. Using the 100-inch Hooker telescope on Mount Wilson in Southern California, Hubble zeroed in on a hazy spiral object in the constellation Andromeda called M31 and proved it was a remote "island universe" in its own right. Today it is known as the Great Galaxy in Andromeda, the closest relative to our own galaxy, the Milky Way.

Hubble's proof involved estimating the distance to Andromeda by using a then-novel astronomical measuring stick called the Cepheid variable method. Cepheid variables are regularly pulsing stars. Their rate of variation directly depends on the amount of energy they put out. The distance of a star can be well gauged by knowing both its energy output rate and its apparent brightness as seen from Earth. For example, if a star puts out considerable amounts of radiation, but still seems dim from Earth, it must be relatively distant. Conversely, if a star has low power output, but still looks bright, it must be close.

One might understand this principle by considering luminous objects on Earth. A bright airplane search light that looks dim must be far away. On the other hand, if the weak flickering bulb of a small flashlight about to expire still looks bright, you must be holding it directly in front of your eyes.

Hubble measured the variation rate of one dozen Cepheids in M31 and also gauged their apparent brightness. From the former data, he calculated their amounts of energy production. Then combining all of this information, he computed their distances, and hence the distance to M31 itself. He found M31 to be about one million light years away, placing it well outside the boundaries of the Milky Way.

He went on to apply the same powerful technique to other seeming nebulae (gas clouds) in space. In case after case, he determined the objects to be stellar communities millions of light years away. Thus, he demonstrated that the universe is composed of many distinct galaxies, a quantity we now believe to number in the tens of billions.

Hubble's second monumental achievement came several years later when he established a relationship between the velocities and distances of twenty-two galaxies that he surveyed. These speeds were assessed by astronomer V. I. Slipher, using the Doppler shifts of radiated galactic light. Recall that light from objects moving away from an observer shifts toward the red end of the spectrum, by an amount related to their speeds. Slipher measured the Doppler shifts of then-known galaxies to establish their velocities relative to us.

In 1929, Hubble, combining Slipher's data with his own distance measurements, reported the discovery of a proportionality now known as Hubble's law: Except for our immediate galactic neighbors, the farther a galaxy away from us, the greater its speed. More-

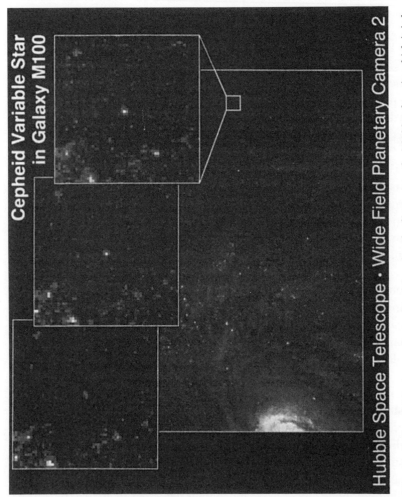

Cepheid Variable Star
in Galaxy M100

Hubble Space Telescope · Wide Field Planetary Camera 2

This Cepheid variable star can be used to determine the distance to the M100 galaxy in which it is located. (Courtesy of NASA.)

over, every other galaxy in the cosmos, excluding our neighbors, is moving away from ours. Thus, in essence he discovered that the universe is flying apart. In the manner brilliantly anticipated by Friedmann's cosmological models, space indeed is expanding.

Critical Tests

Each of Friedmann's universe models begins its life by moving outward from a single point of infinite density. In a brief instant, the cosmos arises from sheer nothingness and begins immediately to expand. We refer to this primordial burst (not strictly an explosion, but rather an outpouring of space itself) as the Big Bang. The expression was coined in the 1940s by astronomer Fred Hoyle as a way of mocking the idea that the everything in the cosmos leapt out of nothingness at the same time. Hoyle found the idea of universal creation to be unsavory; he preferred to think matter and energy are being continuously created in an eternal domain. Hoyle's alternative idea, called the Steady State theory, eventually collapsed under a mounting weight of evidence for the Big Bang model. However, his whimsical name stuck for what is now considered the standard cosmological model.

We live in a still-growing cosmos, approximately fifteen billion years (give or take a few billion) after the Big Bang. Aside from Hub-

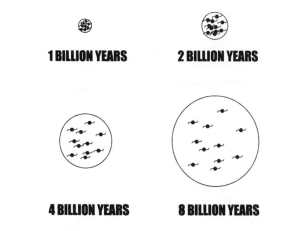

1 BILLION YEARS   **2 BILLION YEARS**

4 BILLION YEARS   **8 BILLION YEARS**

The expansion of the universe causes galaxies to move away from each other over time.

ble's law of galactic recession (movement away from us), vital proof that the universe was once much smaller can be seen in its microwave background radiation.

In 1964, Arno Penzias and Robert Wilson, researchers at Bell Laboratories, accidentally discovered the background "hiss" that comprises the cooled-down, leftover radiation from the Big Bang. While conducting a search for stellar signals with the twenty-foot radio dish at Holmdel, New Jersey, they were disturbed by a persistent background noise. Initially they thought the pigeon droppings—"white dielectric material" they called it—collected on their receiver were causing the hum. They cleaned the dish thoroughly, but still heard the strange buzz.

Shortly thereafter, Princeton theorist Robert Dicke heard about the group's predicament. He immediately realized Penzias and Wilson had discovered radiation present at the time of the Big Bang that had since cooled off because of the universe's expansion. Sure enough, the radio signals found at Holmdel had spectra corresponding to a temperature of three degrees above absolute zero. This closely corresponded with the theorized temperature the background radiation leftover from the Big Bang should be at the present time. For their remarkable discovery, Penzias and Wilson received the Nobel Prize.

In the 1990s, George Smoot extended these findings by using the COBE (Cosmic Background Explorer) satellite to discover ripples in the microwave background, a patchwork of slightly hotter and slightly colder temperatures. These "fluctuations" offer a blueprint of the nascent era of the cosmos, when the radiation observed first burst forth.

According to astrophysical theory, denser regions of the early universe should have left hotter radiation imprints than less dense areas. Smoot demonstrated primordial density discrepancies did indeed leave such telling souvenirs of their earlier configurations. These traces can be found in the patchwork patterns of the cosmic background radiation temperature. The picture Smoot observed corresponds well to what cosmologists believe the early universe was like. Thus Smoot's discovery represents important further confirmation of the Big Bang theory.

Now that the vast majority of scientists are confident the universe originated in a Big Bang, as well as many other details about its

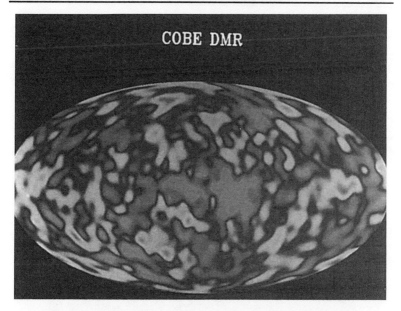

Depicted is a map of the sky's microwave radiation, taken by NASA's Cosmic Background Explorer (COBE) satellite. The light and dark patches indicate slightly hotter and slightly colder regions of the cosmic microwave background radiation. These temperature variations provide insight into the composition of the early universe. (Courtesy of NASA.)

beginnings, many are turning to the intriguing question of what its final stages might be like. Two of the Friedmann models, open and flat, suggest space will keep on expanding forever. However, the third possibility of a closed universe suggests that space, after the current phase of growth, will ultimately shrink back down to a point—and then possibly expand again in a new Big Bang. Obviously not all of these scenarios can be true. Therefore, one might wonder which will be our own destiny.

To determine the fate of the cosmos, one needs to know its current density (its total mass divided by its total volume). If this density is less than or equal to a value called the critical density, estimated to be about 10^{-29} grams per cubic centimeter, we are in an open or flat universe that, in either case, will grow indefinitely. The amount of mass in space will never be great enough to cause it to collapse under its own gravity.

If, in contrast, the current density of the universe is greater than critical, then it is considered closed. Because of the mutual gravitation attraction of its contents, it will someday halt its expansion and begin a stage of contraction—long before the time of heat death. The

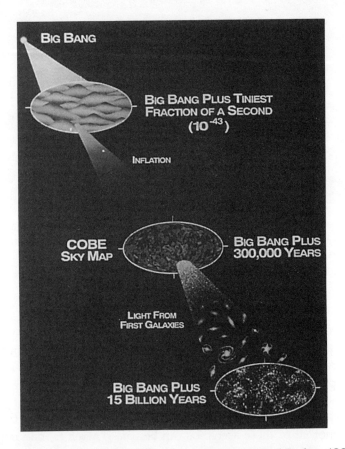

The history of the universe, according to the Cosmic Background Explorer (COBE) satellite. The first oval represents the universe just after the Big Bang. At that time, the cosmos was much smaller and hotter and harbored minute irregularities—seeds of the galaxies we see today. The second oval shows a snapshot of the universe less than half a million years after the Big Bang, as indicated by modern COBE measurements. At that point, the universe had expanded enough and cooled down enough to allow radiation to escape from matter. This primordial light, now much colder, forms the current cosmic microwave background. (Courtesy of NASA.)

complete collapse of the cosmos would likely prevent it from ever reaching a state of maximum disorder. Because of this latter scenario's resemblance to the Big Bang in reverse, it is called the "Big Crunch." For a closed universe, then, a Big Bang expansion lasting billions of years would be followed by a Big Crunch collapse lasting additional billions of years. The cosmos, born in nothingness in a minute embryonic state, would die shrunk down to nothingness as well.

Life in a Crunch

What would it be like for all of space to switch into reverse gear? Except for the telescopic observations of scientists—assuming they still exist billions of years from now—the beginning of a Big Crunch era of the universe would largely go unnoticed. No casual onlooker, living in the hundred millionth century or so, would notice a difference between the expanding and contracting phases of the cosmos. Unlike the mundane experience of changing direction (a car suddenly backing up, for instance), no squealing sounds, unsettling of stomachs, nor any other physical sensations would mark the time when the cosmos halts its expansion and then reverses its course. Superficially, even the sky would look pretty much the same—at least at first. It would still be speckled with stars (though not all the same stars) and galaxies (though not all the same galaxies). Only after contraction takes its course for eons, and the universe significantly collapses, would cosmic temperatures rise and the heavens appear distinctly brighter. But we speak now about the quiet beginning of cosmic contraction, not about its crescendo of an ending.

While no one could note exactly when the universe reverses course, gradually astronomers would start to observe distinct changes in their telescopic readings. Over time they would begin to record marked differences in the light spectra of distant galaxies.

As we've discussed, all galactic light, except for that arriving from our immediate neighbors, is shifted at present toward the red end of the spectrum. Following the Doppler effect, this indicates remote galaxies are moving away from us. However, as space begins to contract, astronomers would start to observe galactic red shifts less and less frequently, and blue shifts more and more often. This bevy of blue shifts would signify more and more galaxies heading toward us. Eventually, when the contraction phase was fully in effect, all galactic spectra would be blue-shifted instead of red-shifted.

Some physicists, such as Stephen Hawking, have speculated that the direction of time's arrow is linked to the dynamics of space. (As we'll see, Hawking has since changed his mind.) Currently, the universe is growing and time is ticking forward. Perhaps this is no coincidence, these theorists have asserted; maybe the Big Bang set the universal clock into motion.

Assuming that expansion and forward movement of the clock are integrally connected, if the universe begins to collapse, might time's direction completely reverse itself? Like a video being reversed on a VCR, might a Big Crunch phase of the universe represent the literal unwinding of the Big Bang?

Imagine how bizarre it would be if time's arrow reverses in a contracting era of the universe. The counterclockwise motions of clocks would set the pace of a strange new existence. People would begin their days by rising groggily out of their beds, unbrushing their teeth, walking backward to hampers filled with dirty clothes, taking out their crumpled clothing, and putting it on. An evening snack, eaten in reverse, would be followed by dinner, and then a backward drive to work. Typical occupations might consist of erasing data for a firm, taking a mop and soiling a floor, lowering rocks into a quarry, or helping pupils forget their multiplication tables. After a hard day's work ended, another reverse commute would take place, followed by a hearty uneating of breakfast. People would remove their clean garments, return them to dresser drawers, and then—feeling refreshed and vibrant—nod off to sleep.

As strange as a time-reversed universe would be, no one would notice the difference. Because everyone would complete their bizarrely reversed actions in synchrony, these gestures would seem to them as prosaic as our forward behavior does today to us. None could step outside of space and objectively observe the backward events transpiring.

Stephen Hawking once argued strongly time would move backward in a big crunch. About a decade ago, he changed his mind. The-

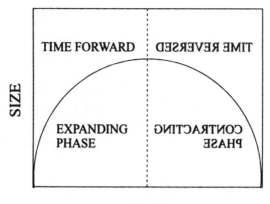

TIME

Some theorists believe the collapse of the cosmos would cause time to reverse course.

oretical calculations that he and several of his colleagues (Don Page of Penn State University and Raymond Laflamme, then at Cambridge) performed led him to discount the possibility of time reversing course in a contracting era. During a Big Crunch—Hawking came to believe—time would simply continue to tick forward in the same manner it does today.

If the universe is open, rather than closed, the question of time reversal would be moot. How might we decide if a Big Crunch lies in our future? According to the models developed by Friedmann, cosmic destiny critically depends on cosmic density. If the cosmos is massive enough, it will eventually collapse; otherwise, it will perpetually grow.

To determine the ultimate fate of the cosmos, astronomers are now trying to measure its density. So far, based on these readings, it looks like the universe is open and will expand forever. Adding up the estimated masses of the galaxies, the total is not great enough ever to reverse expansion and cause contraction. One complicating factor, though, is that there is much invisible (nonradiating) material strewn among the galaxies, as well as within the intergalactic void. Understanding the properties of this dark matter will help settle the question of whether a relatively swift and spectacular Big Crunch or a slow, drawn-out heat death will ultimately represent cosmic destiny.

Dark Forces

Astronomers used to believe the cosmos is solely a collection of brilliant galaxies, grouped into shining stars and other luminous objects. Now they realize visible bodies constitute only a small part of the cosmic terrain. A complete map of the universe would reveal that at least 90 percent, and perhaps as much as 99 percent, of the material in the universe cannot be directly detected. Like an iceberg floating stealthfully in the Arctic, the bulk of the cosmos is invisible. The missing portion is called dark matter.

The first scientific indications of hidden cosmic material were Dutch astronomer Jan Oort's studies of the outer Milky Way in the early 1930s. Oort examined the motions of stars in our galaxy's periphery, and was surprised to find they could not be fully explained by known gravitational forces. The stars were bobbing up and down too much to be experiencing only the tugs of visible matter in the galaxy. His calculations indicated that fully three times the estimated mass of the Milky Way would be required to generate such strong pulls. Clearly something was missing, but what was it?

Swiss–American astronomer Fritz Zwicky conducted a study around the same time as Oort's that reached the similar conclusion that much is unseen in space. Examining a collection of galaxies known as the Coma Cluster, he calculated how much mutual gravitational attraction would be required to keep it from flying apart. To his amazement, fully three hundred times the mass of the visible matter in the cluster would be needed for stability—its glue in a manner of speaking. Since the Coma Cluster is clearly stable, he surmised that most of its material is hidden from view.

Though these observations generated some concern, astronomical interest in dark matter did not blossom until the 1970s, when Vera Rubin and her colleagues at the Carnegie Institution of Washington made a startling discovery. Examining what is called the "rotation curves" of several galaxies—the orbital speeds of the stars that make up each galaxy plotted against their distances from its center—they noticed a strange phenomenon. Stars in the peripheries of galaxies such as Andromeda seemed to move much faster than expected, in each case as if a great quantity of unseen mass was pushing them along with its gravitational force. Rubin's group concluded the haloes (outer regions) of galaxies, once thought to be virtually empty, were brimming with invisible material.

Observation after observation in the 1980s and 1990s verified most of the cosmos is composed of dark matter. Astronomers demonstrated it is present within the domains of galaxies, in intergalactic spaces, and even in the universe's vast empty regions, known as voids. No successful theory of the universe might ignore its mysterious presence.

Vera Rubin, Carnegie professor who provided early evidence of the existence of dark matter. (Courtesy of the Carnegie Institution of Washington and Philip Berming-ham Photography.)

Theories of what constitutes dark matter have abounded since its discovery. The three main contenders are known as MACHOs (Massive Compact Halo Objects), WIMPs (Weakly Interacting Massive Particles), and massive neutrinos. Most researchers today believe dark matter consists of a mixture of at least two of these components.

MACHOs are dim stellar bodies such as red dwarfs, brown dwarfs (stars that do not contain enough fusible material to shine), white dwarfs, neutron stars, and black holes, that lie in the outer regions of galaxies. Until recently, their presence was unproven. However, in 1993 an American–Australian collaboration called the MACHO project, headed by Charles Alcock of Lawrence Livermore National Laboratory and Kim Griest of the University of California at San Diego, used the five-foot diameter telescope at the Mount Stromlo Observatory near Canberra, Australia, to find the first evidence of MACHOs in the Milky Way. They found these hidden objects indirectly—by observing their gravitational bending of the light from remote stars in the Large Magellanic Cloud, a suburb galaxy of our own. Another team of researchers from France, called the EROS project, found similar results at the same time. Since that time, these and other groups have found hundreds of MACHO events.

WIMPs, according to theory, are massive particles pretty much evenly distributed throughout space. The reason they haven't been seen is that they do not readily interact with ordinary types of matter. With exotic names such as axions, photinos, selectrons, smuons, gravitinos, Winos, Zinos, etc.—each term having its own technical meaning in physical theory—WIMP candidates are as strange and abundant as creatures in Australian zoos. They are the cosmological equivalent of duck-billed platypuses, harboring peculiar combinations of properties of known particles. However, unlike the unusual animals that populate the land down under, WIMPs have yet to be captured on camera. They only exist hypothetically, and could conceivably turn out in fact to be more analogous to centaurs and minotaurs than to real denizens of a menagerie.

The third major theory of dark matter concerns the masses of neutrinos. Neutrinos are among the most common elementary particles in the cosmos. They move extremely quickly—at or close to the speed of light—and interact rarely. One of the deep questions of particle physics is whether or not these particles have mass. (Earlier we discussed how this debate pertains to theories of solar dynamics.) If indeed each neutrino has a mass, however small, (as recent experiments seem to indicate) the sum total of these ubiquitous particles would comprise a large chunk of the invisible matter in the universe. Like a backpack loaded down with sand, when something virtually weightless is present in great quantity, the entirety can be quite heavy.

Of the three theoretical constituents of dark matter, MACHOs are the most certain. Yet their presence accounts for only about 20 percent of the gravitational forces detected by Rubin and others. WIMPs, massive neutrinos, or maybe another phenomenon now unknown, must be called in to fill the gap.

Astrophysicists hope a more complete picture of the material content of the universe, including its dark matter, will be developed in coming decades. This will help them determine whether space is closed, open, or flat. Perhaps enough invisible material will be found to "close" the cosmos—meaning that it is dense enough that it will eventually contract. Otherwise, our best portrait of the universe, taking into account its currently estimated density, indicates that it will expand forever.

The Edge of the Universe

In January 1998, researchers presented the most impressive evidence to date that the universe will expand forever. Two teams of astronomers—a group from Lawrence Berkeley Laboratory in California and a group from the Harvard–Smithsonian Center for Astrophysics in Cambridge, Massachusetts—reported measurements of the deceleration parameter (rate of slowing down) of the universe. These values indicate the universe's expansion is presently maintaining such a steady pace that it will never come to a halt.

The method used by these scientists involved recording the speeds and distances of supernova that lie in remote galaxies. Over a number of years they observed a special class of supernova, called Type Ia, known for the consistent level of energy produced in explosions. Like a group of light bulbs, all of the same wattage, Type Ia blasts emit predictable amounts of luminous radiation. Therefore, like the Cepheid variable stars employed by Hubble, Type Ia supernova serve as excellent standard candles: objects whose known energy outputs and measured peak brightnesses can be used to find their distances. The dimmer such a supernova appears, the farther away it is. Employing the Isaac Newton telescope at Las Palmas in the Canary Islands, the Cerro Tololo Observatory in Chile, the Keck Observatory in Hawaii, and other facilities scattered around the world, the teams determined the distances to more than fifty stellar explosions located in remote galaxies. Finding especially distant objects was important because their light was very old—having spanned considerable portions of the cosmos to get to Earth. Therefore these remote explosions provided crucial information about the early universe and its behavior.

One of the Berkeley team's most recent discoveries, for example, was a supernova explosion that took place when the universe was half its current age. Supernova 1997ap, detected on March 5, 1997, with the Cerro Tololo 4-meter (thirteen-foot-diameter) telescope, represents the most distant supernova found to date. Billions of light years away, its observed properties provided critical data about early cosmic expansion.

Whenever astronomers from each team discovered a distant supernova, they needed to act quickly; the blasts they observed faded very rapidly—within days. The demands of their task required them to work steadily, night after night, until the burst was no longer visible. After relaying their data to other groups, they waited for the next explosion.

Saul Perlmutter, leader of the Berkeley team, described to me the frustration of being up all night, for night after night, in windowless observatories, and not being able to enjoy the spectacular mountain scenery that surrounds them:

> Most of the best sites for telescopes are in beautiful parts of the world (Canary Islands, Chile, Hawaii), so you see wonderful landscapes as you drive (or are driven) to the top of the mountain where the telescope is. Then for the next two or three days you are in a windowless control-room, up all night, and asleep most of the day. So you see almost nothing of the wonderful scenery until you are dri-

ving back to the airport. This is particularly a drawback for our supernova work—because we are tracking the rapidly fading supernovae, and usually have to rush back to analyze the data and give instructions to the next telescope that is following the supernovae—so you rarely have time to appreciate the wonderful parts of the world where you are working. Next project, we will have to study some object that just stays put, and doesn't fade![2]

Perlmutter's persistence splendidly paid off. He and his colleagues collected ample data detailing the distances of supernovae around the universe. By combining this information with the velocity values of the galaxies in which the supernovae are located—determined by each galaxy's Doppler shift—he developed the best estimate of the cosmic deceleration parameter to date. He concluded the universe's expansion doesn't seem to be slowing down enough for it to ever reverse course and become a universal contraction. In fact, cosmic expansion may even be speeding up. The Harvard–Smithsonian team announced very similar results.

If indeed the Friedmann models accurately describe cosmic behavior, the universe seems to be open, rather than closed or flat. As Friedmann's open model predicts, the cosmos seems to be growing without limit. Some theorists suggest this phenomenal rate of expansion calls for modification of current theories. They have revived the early suggestion by Einstein—which he later discarded—of the existence of a "cosmological constant" term in his equations of general relativity. If such a term exists, then space has a natural negative pressure, acting somewhat like gravity in reverse, by pushing it ever outward faster and faster. Perlmutter and his colleagues hope that additional supernova surveys will determine whether or not this negative pressure truly exists.

Cosmology will certainly be exciting in years to come, as novel theories attempt to make sense of new astrophysical measurements. As Perlmutter has remarked:

This is the decade in which we will obtain our first large collection of data that will answer many of the fundamental questions of cosmology. We are likely to discover that our first, simplest theories are not quite right (since they were formulated in the absence of much data). Each of the measurements in this field will require very careful (difficult) work to track down possible sources of error. Then the fun will be bringing all of these datasets together and trying to find a theory that fits them.[3]

Though he is justifiably proud of the research goals accomplished by him and his group, Perlmutter is somewhat philosophically disappointed their results suggest the cosmos is destined to expand forever: "I was sort of rooting for a universe that would someday collapse, since an infinitely expanding universe seems like a rather cold, empty—lonely—ending."[4]

Indeed if theorists Fred Adams and Gregory Laughlin are correct, the demise of the universe (assuming it is open, as current evidence seems to indicate) will be a painstaking process drawn out over trillions and trillions of years. Like an actor hamming it up in a campy horror movie, the cosmos will not surrender its life quickly.

Chronicle of the Far Future

In 1997, Fred Adams and Gregory Laughlin published a remarkable research article in the prestigious journal, *Reviews of Modern Physics*. Like most of the papers in that journal, their report, entitled "A dying universe: the long-term fate and evolution of astrophysical objects,"[5] contained little new information, but, rather, summarized state-of-the-art knowledge about the subject. Its merits included its comprehensiveness, its organization, and the strong case that it made for its long range predictions. Focusing on the possibility the cosmos is either flat or open (while also briefly addressing the alternative of a closed universe), it presented a fascinating chronicle that starts at the Big Bang and looks trillions and trillions of years into the future.

Using current scientific understanding, Adams and Laughlin divided the past and future history of a flat or open cosmos into five distinct stages. These eras reflect how the universe functions at those times, particularly its dominant processes for energy production.

The first stage, referred to as the "radiation-dominated era," comprises the initial period after the Big Bang in which the density of radiative energy in the universe was greater than the density of matter. This was the time, well known in astrophysics, in which the universe was hottest.

Next comes the "stelliferous era," starting approximately one million years after the Big Bang, and projected to end tens of trillions of years from now. During that lengthy period—which is the one we currently reside in—the cosmos is dominated by matter, rather than energy. Stellar fusion, particularly of hydrogen into helium, but also of higher elements, provides the main source of power in space. Planetary systems, and the possible forms of life they harbor—including humans—are born, develop, and die throughout that era.

The "degenerate era" that follows will be an age of has-beens. Hot shining suns will turn into cool white dwarfs or even colder stellar states such as brown dwarfs, neutron stars, and black holes. Meanwhile, the baryonic matter within these objects—defined as building-block elementary particles such as protons and neutrons, will engage in an extended process of slow decay. (The ability of protons to decay is a controversial point in modern physics; many theorize such processes occur, but experiments have yet to confirm these theories.) Because of such decay, by the end of the degenerate era, all of these bodies except for the black holes will wither away.

The "black-hole era" will begin, according to Adams and Laughlin, approximately 10^{38} (1 followed by 38 zeroes) years from now. By then, nothing will be left except for black holes. These too will continue to evaporate, slowly radiating away their fuel, in the manner described by Hawking, until there are none remaining.

The last phase of the cosmos, which Adams and Laughlin call the "dark era," will take place approximately 10^{100} (1 followed by one hundred zeroes) years from now. In that fateful era, cosmic entropy will have approached its maximum value. The loneliness of an empty universe will reverberate like an echo in an empty chamber when even the black holes will be long gone.

Once the universe has entered its dying stages, might there be a way for humans and other intelligent beings to survive, assuming they still exist? Almost certainly not, at least according to our current understanding. A few hypothetical methods have been proposed by physicists for intelligence to exist forever. One should note, though, that these are highly speculative.

Freeman Dyson, for example, of the Institute for Advanced Study in Princeton, has speculated that as the cosmos grows colder, intelligent life will develop the capacity to slow down its thought processes and metabolism indefinitely. As the universe grows older and more sluggish, and its temperature lowers, our body temperatures would adjust to these changing conditions. Eventually, our rates of thinking and being would be so drawn out we might operate at conditions close to absolute zero (the lower limit of temperatures).

Russian cosmologist Andrei Linde, currently at Stanford, has proposed another, highly speculative, possible escape route. He sug-

Andrei Linde, Stanford professor who developed prominent cosmological theories. (Courtesy of Andrei Linde.)

gests the warping properties of general relativity allow for the creation of new baby universes. These are minicosmos that break off from the main body of the universe when the mass and energy of a particular region become concentrated enough. Like a rock placed on a stretched sheet, causing it to fold in on itself, enough material prodding at the fabric of the universe (due to a large black hole, for example) might induce it to form new pockets of space. Once a baby universe is formed, Linde projects it would continue to grow, until there is a chance that worlds like Earth might develop within. Travel from our own cosmos to new ones would probably be impossible. However, Linde suggests information might be transmitted to the nascent pockets of space, and the data might eventually be used to continue the legacy of intelligent life elsewhere.

Barring the indefinite slow-down of metabolism as proposed by Dyson, or the extension of life via the production of baby universes as hypothesized by Linde—or some other mechanism, currently unknown—once the cosmos settles into its long quiescence all sentient existence, including the human race and any other intelligent species that might have arisen elsewhere, would be long deceased. As many a sports announcer has said, once the last player has struck out, the game is history.

EPILOGUE

TRIUMPH OVER FEAR

I worry that, especially as the millennium edges nearer,
pseudoscience and superstition will seem year by year
more tempting, the siren song of reason more sonorous
and attractive . . .
The candle flame gutters. Its little pool of light trembles.
The demons begin to stir.

—CARL SAGAN, *"The Demon-Haunted World"*

Around the world, earthquakes, volcanoes and tidal waves periodically ravage human habitats. Pollution, famine, and pestilence frequently imperil our health and livelihood. While nuclear, chemical, and biological warfare stand as a sword of Damocles over our civilization, the possibility of cosmic catastrophe represents the ultimate in horrors—the demise of the universe itself.

With the passing of the year 2000 and the dawn of the third millennium, we toss the two-headed coin of doom and renewal over and over again in our tired minds, wondering which side will ultimately land. Blessed and cursed with the power of imagination, we have the capacity to picture heaven and hell, utopia and armageddon. Perhaps, like Dante's character, we must suffer the tribulations of the Inferno, at least in our thoughts and feelings, before we might enter the gates of Paradise. Or conversely, if the gong of doomsday ever resounds, we will rue our ability to ponder what wonders could have been.

Considering the Pandora's box of horrors that might devastate the Earth at any time, how might we cope with these in a meaningful way and still remain sane? Faced with the specter of nuclear armageddon, destruction of the ozone layer, unchecked disease, massive flooding induced by global warming, life-threatening environmental pollution, cataclysmic new ice ages, cometary collision, stellar explosions, and so on, it would be far too easy either to become unreasonably alarmed or hopelessly jaded. Panic and apathy represent two extreme reactions that hamper our ability to find workable solutions to impending crises.

Demagogues take advantage of these irrational responses to emergencies. Using fiery rhetoric, often divorced from the hard facts of the situation, they feed on the public's desire for a quick and easy solution. They preach the hastiest attempt out of the dilemma, even if it is far from the best, and, in many cases, many even represent the worst. Often their professed answer to a problem is just to follow them blindly, as the Nazis did Hitler. The panicked segment of the populace welcomes these strong leaders, while the apathetic simply go along.

The antidote to the twin maladies of not thinking and not caring lies in science education. With an educated public, the chances of engaged, thoughtful, and meaningful responses to crises are greatly enhanced. The well-read individual is the bane of demagogues. Neither willing to follow the dictates of sectarian movements wholeheartedly, nor to step aside uncaring if they rally their supporters with misleading information, loath to trust wholly in the government, but averse to discount its statements automatically, maintaining faith but eschewing superstition, the educated person makes informed judgments based on the up-to-date facts of the situation.

Science is hardly perfect. From heinous Nazi physician Josef Mengele to doctrinaire Stalinist geneticist Trofim Lysenko, scientists are hardly immune to the misapplication of their skills in the abuse of power. Yet it is instructive to note that scientific abuse generally happens when information is suppressed, opposing views quashed, and dissidents jailed or killed. Conversely, when discussions of ideas are free and open, and the public is educated to participate in them meaningfully, then misapplications are rare.

For example, open discussion of the prospects for human cloning (copying an individual's genetic material and creating another being) has led to public dialogue concerning whether or not such research should be banned. If the public wasn't aware of these possibilities, then scientists could proceed to violate moral boundaries, and later claim ignorance of their research's implications.

Even under the best of circumstances, science makes mistakes, but a healthy scholarly community works hard to spot these errors and correct them. By means of painstaking experimentation and meticulous theoretical research, the scientific model of nature advances toward increasing accuracy and greater utility. Science never achieves perfection; it only represents the best strategy humanity has to offer. As the late Carl Sagan pointed out:

> One of the reasons for its success is that science has built-in, error-correcting machinery at its very heart. Some may consider this an overbroad characterization, but every time we exercise self-criticism, every time we test our ideas against the outside world we are doing science. When we are self-indulgent and uncritical, when we confuse hopes and facts, we slide into pseudoscience and superstition.[1]

Without a doubt, there are areas where science provides no answer. Under certain circumstances, a profound crisis—one leading to doomsday perhaps—might evade solution altogether. Someday, if faced with a global catastrophe, the scientific community might discover to its despair that nothing can be done to prevent the imminent destruction of Earth. (This would almost certainly be the case if the Earth was destroyed suddenly and without warning). Then, and only then, once all sensible avenues have failed and the good fight has unmistakably been lost, should we reluctantly face the dying of the light.

When night falls, but the comforting false oblivion of sleep fails to arrive, our deepest fears—of a far more decisive nothingness—leap into awareness. Like unwanted intruders, lurking at the threshold of doors normally locked, they take advantage of nocturnal openings to come in. Daytime presents illusions of continuity, of comforting locales and familiar faces. But when at night these calming visions have faded into darkness, alarming specters of extinction become all too apparent.

Clock hands turn, but sleep does not come. A horrible premonition grows into a lifeform far more vital than any human. Dread piles up upon dread; despair upon despair. Yet nothing really terrible has happened, has it? Tired minds can sometimes play tricks . . .

A deep rumble, seeming to emerge from the bowels of the Earth, suddenly disturbs us. Might it be the initial tremor of an earthquake? Or the first signal a nuclear bomb has exploded nearby? Perhaps the crashing of an iron asteroid—or an icy comet—has caused our rooms to shake.

What might be our thoughts—our final thoughts—when we realize our nebulous nightmares have sprung forth into vivid reality? Or, conversely, knowing that we have fallen into the bottomless pits of our own worst anxieties. Either way our fate would be as frozen as the chilling air pouring in through shattered windows.

The sharp wails of sirens, the glare of flashing lights, and the pungent smell of blood normally serve as stimulants to action. But no movement seems possible amid the mayhem. The chaos of events has overwhelmed any responses that we might muster.

Another crash. The collapsing walls and ceilings of devastated buildings shatter any remaining sense of hope. Bricks, plaster, stones, and cement mix painfully with flesh. The air grows thick with dust. It's becoming harder and harder just to breathe.

Buried under mountains of rubble, signs of life flowing fast from our veins, what would be our ultimate thoughts? What final images would fill our heads when we face the inevitability of not just our own demise, but the death of our kind as well? What would we think—what *could* we think—when it's the end of the world?

REFERENCES

Preface

1. *Time*, February 25, 1966. As reported in Rose De Wolf, Yesterday's tomorrow, *New York Times Magazine*, December 24, 1995, p. 46.
2. Revelation 22, Verse 18, In *The New Oxford Annotated Bible, Revised Standard Version*. Edited by Herbert G. May and Bruce M. Metzger. (New York: Oxford University Press, 1977), p. 1514.
3. Dylan Thomas, Do Not Go Gentle into That Good Night. In *The Premier Book of Major Poets*, edited by Anita Dore, (Greenwich, Connecticut: Fawcett Publications, 1970), p. 144.

Introduction

1. Norman MacLeod. Reported in What really killed the dinosaurs, *New Scientist*, Vol. 155, No. 2095, August 16, 1997, p. 23.
2. Anthony Allen and Shin Yabushita, Did an impact alone kill the dinosaurs? *Astronomy and Geophysics*, Vol. 38, No. 2, Apr. 1, 1997, p. 15.
3. Reported in What really killed the dinosaurs, *New Scientist*, Vol. 155, No. 2095, 16 August 1997, pp. 25–26.
4. Ibid.
5. Ibid.
6. Dewey McLean, K-T transition greenhouse and embryogenesis dysfunction in the dinosaurian extinctions, *Journal of Geological Education*, Vol. 43, no. 5, Nov. 1, 1995, p. 517.
7. Clark R. Chapman and David Morrison, *Cosmic Catastrophes* (New York: Plenum, 1989), pp. 279–280.
8. Thomas Ahrens and Alan W. Harris, Deflection and fragmentation of near-Earth asteroids. In *Hazards Due to Comets and Asteroids*. Edited by Tom Gehrels. (Tucson: University of Arizona Press, 1994), pp. 897–927.
9. V. A. Simonenko, V. N. Nogin, D. V. Petrov, O. N. Shubin, and Johdale C. Solem, Defending the Earth against impacts from large comets and as-

teroids. In *Hazards Due to Comets and Asteroids*. Edited by Tom Gehrels. (Tucson: University of Arizona Press, 1994), pp. 929–953.

10. B. P. Shafer, M. D. Garcia, R. J. Scammon, C. M. Snell, R. F. Stellingwerf, J. L. Remo, R. A. Managan, and C. E. Rosenkilde, The coupling of energy to comets and asteroids. In *Hazards Due to Comets and Asteroids*. Edited by Tom Gehrels. (Tucson: University of Arizona Press, 1994), pp. 955–1012.

11. Tom Gehrels, personal communication, January 1998.

12. Dylan Thomas, Do Not Go Gentle into That Good Night. In *The Premier Book of Major Poets*, edited by Anita Dore, (Greenwich, Connecticut: Fawcett Publications, 1970), p. 144.

Chapter 1

1. Plato, *Timeaus* 21E–25D. In Eberhard Zangger, *The Flood from Heaven* (London: Sidgwick and Jackson Limited, 1992), p. 21.

2. Philip Freund, *Myths of Creation* (London: W. H. Allen, 1964), p. 10.

3. A more detailed account of this subject appears in: P. Halpern, *The Cyclical Serpent: Prospects for an Ever-Repeating Universe* (New York: Plenum, 1995).

4. Daniel 7: Verses 2–7. In *The New Oxford Annotated Bible, Revised Standard Version*. Edited by Herbert G. May and Bruce M. Metzger. (New York: Oxford University Press, 1977).

5. Daniel 12: Verses 2–3. In *The New Oxford Annotated Bible, Revised Standard Version*. Edited by Herbert G. May and Bruce M. Metzger. (New York: Oxford University Press, 1977).

Chapter 2

1. St. Augustine, *City of God*, translated by Marcus Dods, (New York: Random House, 1950), pp. 718–719.

2. Revelation 20, Verses 1–8. In *The New Oxford Annotated Bible, Revised Standard Version*. Edited by Herbert G. May and Bruce M. Metzger. (New York: Oxford University Press, 1977).

3. Peter 2: Verse 3. In *The New Oxford Annotated Bible, Revised Standard Version*. Edited by Herbert G. May and Bruce M. Metzger. (New York: Oxford University Press, 1977).

4. Norman Cohn, *The Pursuit of the Millennium: Revolutionary Millenarians and Mystical Anarchists of the Middle Ages* (New York: Oxford University Press, 1970), p. 29.

5. Richard A. Landes, *While God Tarried: Disappointed Millennialism and the Making of the Modern West*, unpublished manuscript.

6. R. Glaber, The first millennium. In *Life in the Middle Ages*, edited and

translated by G. G. Coulton. (Cambridge: Cambridge University Press, 1930), pp. 2–3.

7. G. G. Coulton, *Life in the Middle Ages* (Cambridge: Cambridge University Press, 1930), p. 1.

8. Ibid.

9. Richard Erdoes, *A. D. 1000: Living on the Brink of Apocalypse* (San Francisco: Harper & Row, 1988), pp. 1–9.

10. Norman Cohn, *The Pursuit of the Millennium*, pp. 59–60.

11. Laurie Garrett, *The Coming Plague* (New York: Farrar, Strauss and Giroux, 1994), p. 237.

12. Giovanni Boccaccio, *The Decameron.* Translated by G. H. McWilliam. (New York: Penguin Books, 1995), pp. 8–9.

13. Jeffery A. Fisher, *The Plague Makers: How We are Creating Catastrophic New Epidemics—and What We Must Do to Avert Them* (New York: Simon and Schuster, 1994), pp. 22–23.

14. Jeffery A. Fisher, *The Plague Makers*, pp. 151–164.

15. Laurie Garrett, *The Coming Plague*, pp. 103–104.

Chapter 3

1. Valentine Rathbun, *Some Brief Hints of a Religious Scheme*, Salem: S. Hall, 1782, p. 12. Quoted in Edward Deming Andrews, *The People Called Shakers*, New York: Dover, 1963, p. 28.

2. William Miller, *Miller's Works.* Edited by Joshua V. Himes. (Boston: Joshua V. Himes, 1842), Vol. 1, p. 9. Cited by Wayne R. Judd, William Miller: disappointed prophet, in *The Disappointed: Miller and Millenarianism in the Nineteenth Century.* Edited by Ronald L. Numbers and Jonathan M. Butler. (Knoxville: The University of Tennessee Press, 1993), p. 18.

3. Wayne R. Judd, William Miller: disappointed prophet, in *The Disappointed: Miller and Millenarianism in the Nineteenth Century.* Edited by Ronald L. Numbers and Jonathan M. Butler. (Knoxville: The University of Tennessee Press, 1993), p. 23.

4. Ibid., p. 34.

5. Lawrence Foster, Had prophecy failed? in *The Disappointed: Miller and Millenarianism in the Nineteenth Century.* Edited by Ronald L. Numbers and Jonathan M. Butler. (Knoxville: The University of Tennessee Press, 1993), pp. 179–181.

6. Melvin D. Curry, *Cults and Nonconventional Religious Groups: Jehovah's Witnesses: The Millenarian World of the Watch Tower,* (New York: Garland Publishing, 1992), p. 89.

7. Keith Harrary, The truth about Jonestown, *Religious Cults in America.* Edited by Robert Emmet Long. (New York: H. W. Wilson Co., 1994), p. 13.

Chapter 4

1. Stephen Schneider, *Global Warming: Are We Entering the Greenhouse Century?* (San Francisco: Sierra Club Books, 1989), pp. 50–51.
2. Ibid.
3. Richard B. Stothers, The Great Tambora Eruption of 1816 and Its Aftermath, *Science,* Vol. 224, No. 4654, (June 1984), pp. 1191–1197.

Chapter 5

1. Jamie Sayen, *Einstein in America: The Scientist's Conscience in the Age of Hitler and Hiroshima,* (New York: Crown Publishers, 1985), p. 117.
2. Albert Einstein, *Einstein on Peace.* Edited by Otto Nathan and Heinz Norden. (New York: Simon and Schuster, 1960), p. 290. Reported in Jamie Sayen, *Einstein in America: The Scientist's Conscience in the Age of Hitler and Hiroshima,* (New York: Crown Publishers, 1985), p. 118.
3. Jamie Sayen, *Einstein in America,* p. 122.
4. Helen Caldicott, *Missile Envy: The Arms Race and Nuclear War,* (New York: Bantam Books, 1986), pp. 10–11.

Chapter 6

1. Melinda Kimble. Reported in Mark Jaffe, A challenge of global proportions, *The Philadelphia Inquirer,* Nov. 30, 1997, p. A21.

Chapter 7

1. David E. Fisher, *Fire and Ice: The Greenhouse Effect, Ozone Depletion and Nuclear Winter,* (New York: Harper and Row, 1990), p. 4.
2. David Morrison, Clark Chapman, and Paul Slovic, The impact hazard. In *Hazards Due to Comets and Asteroids.* Edited by Tom Gehrels. (Tucson: University of Arizona Press, 1994), pp. 59–75.
3. Clark R. Chapman and David Morrison, *Cosmic Catastrophes,* pp. 100–101.
4. Fred L. Whipple, *The Mystery of Comets,* (Washington: Smithsonian Institution Press, 1985), p. 245.
5. Morrison, Chapman, and Slovic, The impact hazard, pp. 61–76.
6. David Crawford. Interviewed by Paul Hoversten, Comet could equal millions of atom bombs, *USA Today,* April 16, 1997, Science, p. 1.
7. Morrison, Chapman, and Slovic, The impact hazard, p. 59.
8. John S. Lewis, *Rain of Iron and Ice: The Very Real Threat of Comet and Asteroid Bombardment,* (New York: Addison-Wesley, 1996), p. 185.

9. Carolyn Shoemaker interviewed. In Rosie Mestel, Night of the strangest comet. *New Scientist*, Vol. 143, No. 1933. (1994), p. 24.

10. Tom Gehrels, *On the Glassy Sea: An Astronomer's Journey*, (New York: American Institute of Physics, 1988), pp. 184–185.

11. Arthur C. Clarke, *Rendezvous with Rama*, New York: Ballantine Books, 1973, pp. 2–3.

12. John S. Lewis, *Rain of Iron and Ice*, p. 89.

13. V. A. Simonenko, V. N. Nogin, D. V. Petrov, O. N. Shubin, and Johdale C. Solem, Defending the Earth against impacts from large comets and asteroids, In *Hazards Due to Comets and Asteroids*. Edited by Tom Gehrels. (Tucson: University of Arizona Press, 1994), pp. 929–953.

14. Stewart Nozette. Reported in Jeff Hecht, Pentagon hot shots take aim at asteroids, *New Scientist*, Vol. 149, No. 2022, 23 March 1996, p. 12.

15. Jeff Hecht, Asteroid spared, *New Scientist*, Vol. 156, No. 2105, 25 October 1997, p. 5.

Chapter 8

1. John Updike, Cosmic gall, In *Telephone Poles and Other Poems*, (New York: Knopf, 1963), p. 1.

2. H. G. Wells, *The Time Machine*. In *Seven Science Fiction Novels of H.G. Wells*, (New York: Dover Publications, 1934), p. 70.

Chapter 9

1. See Paul Halpern, *The Cyclical Serpent*, pp. 221–266.

2. Saul Perlmutter, personal communication, January 1998.

3. Ibid.

4. Ibid.

5. Fred C. Adams and Gregory Laughlin, A dying universe: the long-term fate and evolution of astrophysical objects, *Reviews of Modern Physics*, Vol. 69, No. 2, April 1997, pp. 337–370.

Epilogue

1. Carl Sagan, *The Demon-Haunted World: Science as a Candle in the Dark* (New York: Random House, 1995), p. 27.

RELATED READING

The following is a list of general and technical readings related to religious and scientific visions of the end of the world. Technical references are indicated with asterisks.

Introduction

*Allen, Anthony and Yabushita, Shin, Did an impact alone kill the dinosaurs? *Astronomy and Geophysics*, Vol. 38, No. 2 (Apr. 1, 1997): 1.

Chapman, Clark R. and Morrison, David, *Cosmic Catastrophes* (New York: Plenum, 1989).

Gehrels, Tom, *Hazards Due to Comets and Asteroids*. (Tucson: University of Arizona Press, 1994).

*McLean, Dewey, K-T Transition greenhouse and embryogenesis dysfunction in the dinosaurian extinctions, *Journal of Geological Education*, Vol. 43, No. 5 (Nov. 1, 1995): 1.

Chapter 1

Boyce, Mary, *Zoroastrians: Their Religious Beliefs and Practices* (Boston: Routledge and Kegan Paul, 1979).

Eliade, Mircea, *Cosmos and History: The Myth of the Eternal Return* (New York: Harper and Row, 1959).

Freund, Philip, *Myths of Creation* (London: W.H. Allen, 1964).

Halpern, Paul, *The Cyclical Serpent: Prospects for an Ever-Repeating Universe* (New York: Plenum, 1995).

Halpern, Paul, *Time Journeys: A Search for Cosmic Destiny and Meaning* (New York: McGraw-Hill: 1990).

Heidel, Alexander, *The Gilgamesh Epic and Old Testament Parallels* (Chicago, University of Chicago Press, 1949).

Huggett, Richard, *Cataclysms and Earth History* (New York: Oxford University Press, 1989).

Lewis, Jack Pearl, *A Study of the Interpretation of Noah and the Flood in Jewish and Christian Literature* (Leiden: E. J. Brill, 1968).

Sollberger, Edmund, *The Babylonian Legend of the Flood* (London: British Museum, 1962).

Zangger, Eberhard, *The Flood from Heaven* (London: Sidgwick and Jackson Limited, 1992).

Chapter 2

Augustine, *City of God*, translated by Marcus Dods. (New York: Random House, 1950).

Boccaccio, Giovanni, *The Decameron*, Translated by G. H. McWilliam. (New York: Penguin Books, 1995).

Cohn, Norman, *The Pursuit of the Millennium: Revolutionary Millenarians and Mystical Anarchists of the Middle Ages* (New York: Oxford University Press, 1970).

Coulton, G.G., *Life in the Middle Ages* (Cambridge: Cambridge University Press, 1930).

Emmerson, Richard K., *The Apocalyptic Imagination in Medieval Literature* (Philadelphia: University of Pennsylvania Press, 1992).

Erdoes, Richard, *A.D. 1000: Living on the Brink of Apocalypse* (San Francisco: Harper & Row, 1988).

Fisher, Jeffery, A., *The Plague Makers: How We are Creating Catastrophic New Epidemics—and What We Must Do to Avert Them* (New York: Simon and Schuster, 1994).

Garrett, Laurie, *The Coming Plague* (New York: Farrar, Strauss and Giroux, 1994).

Levy, Stuart B., *The Antibiotic Paradox: How Miracle Drugs are Destroying the Miracle* (New York: Plenum, 1992).

Long, Robert E., *Religious Cults in America* (New York: H.W. Wilson, 1994).

McGinn, Bernard, *Visions of the End: Apocalyptic Traditions in the Middle Ages* (New York: Columbia University Press, 1979).

Chapter 3

Andrews, Edward Deming, *The People Called Shakers* (New York: Dover, 1963), p. 28.

Curry, Melvin D., *Jehovah's Witnesses: The Millenarian World of the Watch Tower* (New York: Garland Publishing, 1992).

Daniels, Ted, *Millennialism: An International Bibliography* (New York: Garland Publishers, 1992).

Harrary, Keith, The truth about Jonestown, *Religious Cults in America,* edited by Robert Emmet Long. (New York: H.W. Wilson Co., 1994).

Lewicki, Zbigniew, *The Bang and the Whimper: Apocalypse and Entropy in American Literature* (Westport, Conn.: Greenwood Press, 1984).

Numbers, Ronald L. and Butler, Jonathan M., *Millerism and Millenarianism in the Nineteenth Century* (Knoxville: University of Tennessee Press, 1993).

O'Leary, Stephen D., *Arguing the Apocalypse: A Theory of Millennial Rhetoric* (New York: Oxford University Press, 1994).

Penton, M. James, *Apocalypse Delayed: The Story of Jehovah's Witnesses* (Toronto: University of Toronto Press, 1985).

Wright, Stuart A., *Armageddon in Waco: Critical Perspectives on the Branch Davidian Conflict* (Chicago: University of Chicago Press, 1995).

Chapter 4

Gribbin, John R., *Children of the Ice: Climate and Human Origins* (Oxford: Basil Blackwell, 1990).

Pielou, E. C., *After the Ice Age: The Return of Life to Glaciated North America* (Chicago: University of Chicago Press, 1991).

Schneider, Stephen, *Global Warming: Are We Entering the Greenhouse Century?* (San Francisco: Sierra Club Books, 1989).

*Stothers, Richard B., The Great Tambora eruption of 1816 and its aftermath, *Science*, Vol. 224, No. 4654 (June 1984): 1191.

Chapter 5

Beres, Louis Rene, *Apocalypse: Nuclear Catastrophe in World Politics* (Chicago: University of Chicago Press, 1980).

Caldicott, Helen, *Missile Envy: The Arms Race and Nuclear War,* (New York: Bantam Books, 1986).

Einstein, Albert, *Einstein on Peace,* edited by Otto Nathan and Heinz Norden. (New York: Simon and Schuster, 1960).

Fisher, David E., *Fire and Ice: The Greenhouse Effect, Ozone Depletion and Nuclear Winter,* (New York: Harper and Row, 1990).

Openshaw, Stan, Steadman, Philip, and Greene, Owen, *Doomsday: Britain after Nuclear Attack* (Oxford: B. Blackwell, 1983).

Plate, Thomas Gordon, *Understanding Doomsday: A Guide to the Arms Race for Hawks, Doves, and People* (New York: Simon and Schuster, 1971).

Sayen, Jamie, *Einstein in America: The Scientist's Conscience in the Age of Hitler and Hiroshima* (New York: Crown Publishers, 1985).

Chapter 6

Dotto, Lydia, *The Ozone War* (Garden City, N.Y.: Doubleday, 1978).

Fisher, David E., *Fire and Ice: The Greenhouse Effect, Ozone Depletion and Nuclear Winter* (New York: Harper and Row, 1990).

McKibben, Bill, *The End of Nature* (New York: Random House, 1989).

Nilsson, Annika, *Greenhouse Earth* (New York: J. Wiley, 1992).

Parsons, Michael L., *Global Warming: The Truth Behind the Myth* (New York: Insight Books, 1995).

Roan, Sharon, *Ozone crisis: The 15-year Evolution of a Sudden Global Emergency* (New York: Wiley, 1989).

Chapter 7

Chapman, Clark R. and Morrison, David, *Cosmic Catastrophes* (New York: Plenum, 1989).

Desonie, Dana, *Cosmic Collisions* (New York: Henry Holt & Co., 1996).

*Gehrels, Tom, *Hazards Due to Comets and Asteroids* (Tucson: University of Arizona Press, 1994).

Gehrels, Tom, *On the Glassy Sea: An Astronomer's Journey* (New York: American Institute of Physics, 1988).

Goldsmith, Donald, *Nemesis: The Death-Star and Other Theories of Mass Extinction* (New York: Walker, 1985).

Lewis, John S., *Rain of Iron and Ice: The Very Real Threat of Comet and Asteroid Bombardment* (New York: Addison-Wesley, 1996).

Mestel, Rosie, Night of the strangest comet. *New Scientist* Vol. 143, No. 1933. (1994): 24.

Whipple, Fred L., *The Mystery of Comets* (Washington: Smithsonian Institution Press, 1985).

Chapter 8

Friedman, Herbert, *The Astronomer's Universe: Stars, Galaxies and Cosmos* (New York: Ballantine Books, 1990).

Goldsmith, Donald, *Supernova: The Violent Death of a Star* (New York: Oxford University Press, 1990).

Jastrow, Robert, *Until the Sun Dies* (New York: Norton, 1977).

Macvey, John W., *Where Will We Go When the Sun Dies?* (New York: Stein and Day, 1983).

Sagan, Carl, *Cosmos* (New York: Ballantine Books, 1980).

Wentzel, Donat G., *The Restless Sun* (Washington: Smithsonian Institution Press, 1989).

Chapter 9

*Adams, Fred C. and Laughlin, Gregory, A dying universe: The long-term fate and evolution of astrophysical objects, *Reviews of Modern Physics*, Vol. 69, No. 2 (April 1997): 337.

Bartusiak, Marcia, *Through A Universe Darkly: A Cosmic Tale of Ancient Ethers, Dark Matter and the Fate of the Universe* (New York: HarperCollins, 1993).

Bartusiak, Marcia, *Thursday's Universe* (New York: Times Books, 1986).

Branch, David, Density and destiny, *Nature*, Vol. 391, No. 1 (January 1, 1998): 23.

Close, Frank E., *Apocalypse When?: Cosmic Catastrophe and the Fate of the Universe* (New York: Morrow, 1988).

*Dyson, Freeman, Time without end: Physics and biology in an open universe. *Reviews of Modern Physics*, Vol. 51, No. 7 (July 1979): 447.

Gribbin, John, *The Omega Point: The Search for the Missing Mass and the Ultimate Fate of the Universe* (New York: Bantam Books, 1988).

Halpern, Paul, *The Cyclical Serpent: Prospects for an Ever-Repeating Universe* (New York: Plenum, 1995).

Halpern, Paul, *The Structure of the Universe* (New York: Henry Holt, 1997).

Hawking, Stephen, *A Brief History of Time: From the Big Bang to Black Holes* (New York: Bantam Books, 1988).

*Hawking, Stephen, Laflamme, R., and Lyons, G. W., Origin of time asymmetry. *Physical Review* D15 Vol. 47 (June, 1993) 5342.

*Linde, Andrei, Life after inflation. *Physics Letters B*, Vol. 211 (August 1988): 29.

Parker, Barry, *Invisible Matter and the Fate of the Universe* (New York: Plenum, 1989).

*Perlmutter, Saul, et. al., Discovery of a supernova explosion at half the age of the universe, *Nature*, Vol. 391, No. 1 (January 1, 1998): 51.

Silk, Joseph, *The Big Bang: The Creation and Evolution of the Universe* (San Francisco: W. H. Freeman, 1980).

Smoot, George, and Davidson, Kay, *Wrinkles in Time* (New York: William Morrow, 1993).

*Trimble, Virginia, Existence and nature of dark matter in the universe. In *The Early Universe: Reprints*, edited by Edward Kolb and Michael Turner (New York: Addison-Wesley, 1988).

Epilogue

Leslie, John, *The End of the World: The Science and Ethics of Human Extinction* (New York: Routledge, 1996).

Sagan, Carl, *The Demon-Haunted World: Science as a Candle in the Dark* (New York: Random House, 1995).

APOCALYPSE-RELATED WEB SITES

The following is a sampling of World Wide Web sites pertaining to apocalyptic ideas. Please note that the Web is not professionally refereed, and some of the information contained in these sources could be inaccurate.

BULLETIN OF THE ATOMIC SCIENTISTS
http://www.bullatomsci.org

CENTER FOR MILLENNIAL STUDIES
http://www.mille.org

ASTEROID AND COMET IMPACT HAZARDS
http://impact.arc.nasa.gov/index.html

COMET HALE-BOPP INFORMATION
http://www.halebopp.com

THE EPA GLOBAL WARMING SITE
http://www.epa.gov/globalwarming/

THE EPA OZONE DEPLETION SITE
http://www.epa.gov/ozone/

GLOBAL WARMING CENTRAL
http://www.law.pace.edu/env/energy/globalwarming.html

SPACEWATCH PROJECT
http://pirlwww.lpl.arizona.edu/spacewatch/

INDEX